Studies in Computational Intelligence 440

Editor-in-Chief

Prof. Janusz Kacprzyk
Systems Research Institute
Polish Academy of Sciences
ul. Newelska 6
01-447 Warsaw
Poland
E-mail: kacprzyk@ibspan.waw.pl

For further volumes:
http://www.springer.com/series/7092

Aleksander Nawrat, Krzysztof Simek,
and Andrzej Świerniak (Eds.)

Advanced Technologies for Intelligent Systems of National Border Security

Springer

Editors
Prof. Aleksander Nawrat
Institute of Automatic Control
Silesian University of Technology
Gliwice
Poland

Prof. Andrzej Świerniak
Institute of Automatic Control
Silesian University of Technology
Gliwice
Poland

Dr. Krzysztof Simek
Institute of Automatic Control
Silesian University of Technology
Gliwice
Poland

ISSN 1860-949X e-ISSN 1860-9503
ISBN 978-3-642-31664-7 e-ISBN 978-3-642-31665-4
DOI 10.1007/978-3-642-31665-4
Springer Heidelberg New York Dordrecht London

Library of Congress Control Number: 2012941231

Printed on acid-free paper

Springer is part of Springer Science+Business Media (www.springer.com)

"Although you can probably hear me, you are not listening..."

To our friends

Preface

The book was motivated by a great number of tasks and challenges placed upon the current administration and various types of uniformed services. A natural, obligatory task is to protect one's own borders from direct and indirect threats. Nowadays, in times of peace, direct threats subside with indirect threats like, for instance, illegal immigration and clandestine transportation of illegal substances and materials such as drugs, weapons and explosive materials. Up until now, conventional methods have been used for detection and thwarting such threats. However rapid progress of development of advanced sensory devices, significant increase of computing power and development of drones technology give almost unlimited possibilities to fight various types of pathologies affecting our society. The book shall address multiple issues on advanced innovative methods of multi-level control of both ground (UGVs) and aerial drones (UAVs). Those objects combined with innovative algorithms become autonomous objects capable of patrolling chosen borderland areas by themselves and automatically inform the operator of the system about potential place of detection of a specific incident. This is achieved by using sophisticated methods of generation of non-collision trajectory for those types of objects and enabling automatic integration of both ground and aerial unmanned vehicles. In order to allow detection of specific events innovative methods of image fusion of heterogeneous camera systems had been developed. Source video streams taken into consideration come from high quality visual light cameras, noctovision cameras and from thermovision cameras. Combining such equipment and previously mentioned algorithms allows to effectively detect threats regardless of the time of day and night, weather conditions or random events such as smoke during a fire. Practical examples of application of such systems shall be presented and described in the following chapters. Obviously, it is natural to propose innovative solutions in the field of mechanical constructions for elimination of the impact of vibrations and other undesirable factors affecting the quality of obtained information streams.

The topics included in this book also cover presentation of complete information and communication technology (ICT) systems capable of control, observation and detection of various types of incidents and threats. This book is not only intended for officers and border services. It can be treated as a valuable source of information for constructors and developers of such solutions for uniformed services. Scientists and researchers involved in computer vision, image processing, data fusion, control algorithms or IC can find many valuable suggestions and solutions. Multiple challenges for such systems are also presented. Moreover students of higher years of study in the fields of automation, computer science, robotics or mechanical engineering can benefit greatly. The authors encourage to actively explore the following chapters of the book and, in case of questions or concerns, invite you to an open and frank discussion of these matters.

Gliwice
May 2012

Aleksander Nawrat
Krzysztof Simek
Andrzej Świerniak

Acknowledgements

The research presented in the book has been done within a framework of the project O R 00 0132 12: *"Design and implementation of innovative unmanned mobile platforms to monitor the state border"* and have been supported by The Polish National Center for Research and Development funds in years 2010 – 2012.

Contents

Part III Sensoric and Vision Based Information for UAVs

Part IV Construction and Implementation of the Base Station and Software for Autonomous Objects

List of Contributors

Artur Babiarz
Silesian University of Technology, Institute of Automatic Control, Akademicka 16, 44-101 Gliwice, Poland
e-mail: artur.babiarz@polsl.pl

Małgorzata Bach
Silesian University of Technology, Institute of Computer Science, Akademicka 16, 44-101 Gliwice, Poland
e-mail: Malgorzata.Bach@polsl.pl

Damian Bereska
Silesian University of Technology, Institute of Automatic Control, Akademicka 16, 44-101 Gliwice, Poland
e-mail: damian.bereska@polsl.pl

Marcin Błaszczok
WASKO S.A., Berbeckiego 6, 44-100 Gliwice, Poland
e-mail: m.blaszczok@wasko.pl

Adam Czornik
Silesian University of Technology, Institute of Automatic Control, Akademicka 16, 44-101 Gliwice, Poland
e-mail: Adam.Czornik@polsl.pl

Krzysztof Daniec
Silesian University of Technology, Institute of Automatic Control, Akademicka 16, 44-101 Gliwice, Poland
e-mail: krzysztof.daniec@polsl.pl

Piotr Demski
Silesian University of Technology, Institute of Automatic Control, Akademicka 16, 44-101 Gliwice, Poland
e-mail: piotr.demski@gmail.com

Adam Duszeńko
Silesian University of Technology, Institute of Computer Science, Akademicka 16, 44-101 Gliwice, Poland
e-mail: Adam.Duszenko@polsl.pl

Marek Fischer
WASKO S.A., Berbeckiego 6, 44-100 Gliwice, Poland
e-mail: m.fischer@wasko.pl

Sławomir Fraś
Silesian University of Technology, Institute of Automatic Control, Akademicka 16, 44-101 Gliwice, Poland
e-mail: boskidialer@gmail.com

Adam Gałuszka
Silesian University of Technology, Institute of Automatic Control, Akademicka 16, 44-101 Gliwice, Poland
e-mail: adam.galuszka@polsl.pl

Arkadiusz Gwóźdź
WASKO S.A., Berbeckiego 6, 44-100 Gliwice, Poland
e-mail: a.gwozdz@wasko.pl

Jarosław Homa
Silesian University of Technology, Institute of Automatic Control, Akademicka 16, 44-101 Gliwice, Poland
e-mail: jaroslaw.homa@polsl.pl

Łukasz Hoppe
WASKO S.A., Berbeckiego 6, 44-100 Gliwice, Poland
e-mail: l.hoppe@wasko.pl

Witold Ilewicz
Silesian University of Technology, Institute of Automatic Control, Akademicka 16, 44-101 Gliwice, Poland
e-mail: witold.ilewicz@polsl.pl

Krzysztof Jaskot
Silesian University of Technology, Institute of Automatic Control, Akademicka 16, 44-101 Gliwice, Poland
e-mail: krzysztof.jaskot@polsl.pl

Karol Jędrasiak
Silesian University of Technology, Institute of Automatic Control, Akademicka 16, 44-101 Gliwice, Poland
e-mail: karol.jedrasiak@polsl.pl

Przemysław Koralewicz
ABB Research Center, Starowiślna 13A, 31-038 Krakow, Poland
e-mail: przemyslaw.koralewicz@pl.abb.com

Roman Koteras
Silesian University of Technology, Institute of Automatic Control, Akademicka 16, 44-101 Gliwice, Poland
e-mail: roman.koteras@polsl.pl

Tomasz Krawczyk
WASKO S.A., Berbeckiego 6, 44-100 Gliwice, Poland
e-mail: t.krawczyk@wasko.pl

Beata Krupanek
Silesian University of Technology, Institute of Measurement Science, Electronics and Control, Akademicka 10, 44-101 Gliwice, Poland,
e-mail: Beata.Krupanek@polsl.pl

Zygmunt Kuś
Silesian University of Technology, Institute of Automatic Control, Akademicka 16, 44-101 Gliwice, Poland
e-mail: zygmunt.kus@polsl.pl

Martin Mellado
Instituto de Automatica e Informatica Industrial, Universidad Politecnica de Valencia, Camino de Vera 14, 46022 Valencia, Espana
e-mail: martin@ai2.upv.es

MichałA. Mikulski
Silesian University of Technology, Institute of Automatic Control, Akademicka 16, 44-101 Gliwice, Poland
e-mail: michal.mikulski@polsl.pl

Aleksander Nawrat
Silesian University of Technology, Institute of Automatic Control, Akademicka 16, 44-101 Gliwice, Poland
e-mail: Aleksander.Nawrat@polsl.pl

MichałNiezabitowski
Silesian University of Technology, Institute of Automatic Control, Akademicka 16, 44-101 Gliwice, Poland
e-mail: Michal.Niezabitowski@polsl.pl

Marcin Pacholczyk
Silesian University of Technology, Institute of Automatic Control, Akademicka 16, 44-101 Gliwice, Poland
e-mail: marcin.pacholczyk@polsl.pl

Krzysztof Skórka
WASKO S.A., Berbeckiego 6, 44-100 Gliwice, Poland
e-mail: k.skorka@wasko.pl

Krzysztof Skrzypczyk
Silesian University of Technology, Institute of Automatic Control, Akademicka 16, 44-101 Gliwice, Poland
e-mail: krzysztof.skrzypczyk@polsl.pl

Marcin Śleziona
WASKO S.A., Berbeckiego 6, 44-100 Gliwice, Poland
e-mail: m.sleziona@wasko.pl

Tadeusz Topór-Kamiński
Silesian University of Technology, Institute of Measurement Science, Electronics and Control, Akademicka 10, 44-101 Gliwice, Poland
e-mail: Tadeusz.Topor-Kaminski@polsl.pl

Tobiasz Warmuz
Silesian University of Technology, Institute of Automatic Control, Akademicka 16, 44-101 Gliwice, Poland
e-mail: tobiasz.warmuz@gmail.com

Łukasz Wąsek
WASKO S.A., Berbeckiego 6, 44-100 Gliwice, Poland
e-mail: l.wasek@wasko.pl

Aleksandra Werner
Silesian University of Technology, Institute of Computer Science, Akademicka 16, 44-101 Gliwice, Poland
e-mail: Aleksandra.Werner@polsl.pl

Katarzyna Wydmańska
Silesian University of Technology, Institute of Automatic Control, Akademicka 16, 44-101 Gliwice, Poland
e-mail: katarzyna.wydmanska@polsl.pl

Part I
Design of Control Algorithms for UAV Objects

There are multiple types of unmanned vehicles adapted to various operation conditions. Depending on the type of terrain, there are objects adjusted to water, ground and air environment. Regardless of the type of the object, they are mostly used for continuous monitoring and patrolling of some areas. Trained personnel and autonomous robots have to cooperate in order to cover with their monitoring perception of vast areas of border lands. At the moment a single operator due to limited perception and difficulty of commanding unmanned vehicles is capable of operating only one unit at a time. In order to maximize the ease of use and, therefore, allow using more than one unit by single operator it is required to implement as much autonomous functionalities of the objects as possible. For instance, automatic stabilization of UAVs or automatic landing and taking-off. Modern control algorithms utilize information from all the sensors mounted to the UAV such as: IMU (accelerometers, gyroscopes, magnetometers), altimeters, GPS and air pressure sensor. The collected data are in each time unit selected and filtered with the use of the probabilistic optimization (especially particle swarm optimization) and classic approaches.

The presented articles show that multilevel control algorithms allow to use UAVs by a person without pilot experience. An example of helicopter flight to a set target location without aid of the operator is presented. It is a significant step forward on the way to create fully autonomous UAV capable of cooperating with human personnel. However, patrolling borderlands does not only require observation from the above but It is also essential for air units to cooperate with the ground support. As an example of such ground support an autonomous mobile platform that can operate in a human-made environment is presented.

Helicopter Control Algorithms from the Set Orientation to the Set Geographical Location

Zygmunt Kuś and Sławomir Fraś

Abstract. The aim of the paper was to develop helicopter control algorithms. On the basis of the results it is possible to control the unmanned helicopter by a person without pilot experience. The suggested solution is based on multilevel control. It allowed to have an effect on helicopter relocation due to setting a requested trajectory. The knowledge mentioned in the literature which concerns the behavior of the helicopter is applied to the synthesis of control system . What is more, the identification of the certain helicopter model elements was made herein. The paper presents the example of the helicopter flight to a set target location. Furthermore, the example provided in the study confirms the correct control system behavior. The possibility of introducing the set location in the coordinate system, which is connected with the ground, allows to reach the set geographical location by the helicopter in an autopilot mode.

Keywords: Unmanned Aerial Vehicle, helicopter model, step response, identification, multilevel control.

1 Introduction

The specific type of an Unmanned Aerial Vehicle (UAV) is considered in the paper. It is a 'miniature version' of a conventional helicopter with a main rotor and tail rotor. The helicopter model which was programmed in C language was used as a controlled plant. Helicopter properties are accurately described by this model. Control plant operation as a rigid body (rotational motion) and particle (translational

Zygmunt Kuś · Sławomir Fraś
Silesian University of Technology, Institute of Automatic Control,
Akademicka 16, 44-101 Gliwice, Poland
e-mail: zygmunt.kus@polsl.pl,
boskidialer@gmail.com

A. Nawrat et al. (Eds.): Advanced Technologies for Intelligent Systems, SCI 440, pp. 3–13.
springerlink.com

motion) are faithfully represented. It also reflects most of important real helicopter behaviors. The method which was proposed by the authors undertakes the development of four control levels.

2 Helicopter Model

The helicopter as the whole may be considered as a system with four inputs and certain number of outputs. This is due to the fact that the number of the outputs depends on a number of the measured signals which are used to the helicopter control.

Figure 1 presents input signals and output signals of the helicopter model.

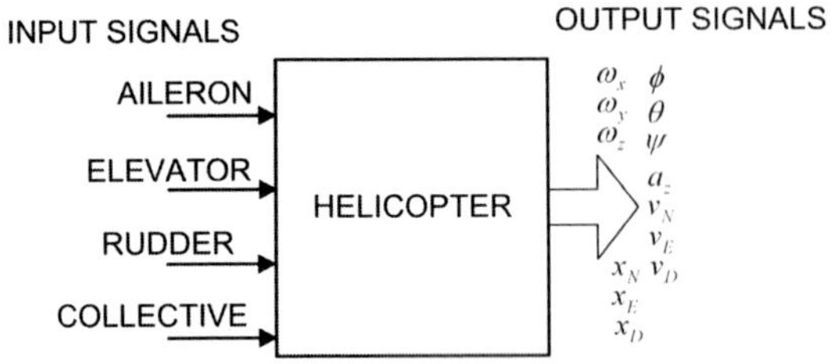

Fig. 1 Input and output signals of the helicopter model

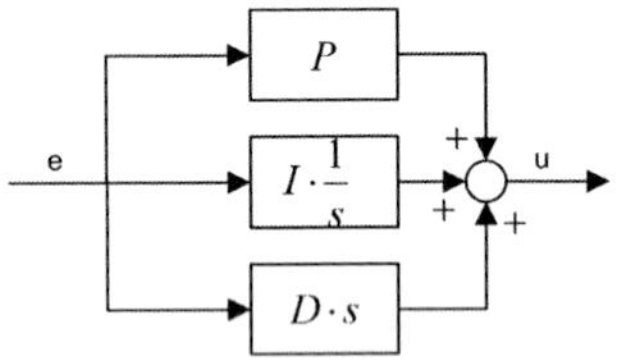

Fig. 2 PID controller

We can define three coordinate axes which are connected with the helicopter, namely longitudinal X, lateral Y and vertical Z. There are defined three angular velocities around X, Y, Z axes - ω_x, ω_y, ω_z.

Morever, we can define three coordinate axes which are connected with the ground: - these are North - x, East - y and vertical Down - z.

We assume the following signal notation:

Input signals:
AILERON $= U_{AI}$, ELEVATOR $= U_{EL}$, RUDDER $= U_{RU}$, COLLECTIVE $= U_{CO}$
Output signals:
Angular velocities: ω_x, ω_y, ω_z. The angles of rotation: Φ, Θ, Ψ. Helicopter linear accelerations: a_x, a_y, a_z. (x corresponds to North, y corresponds to East, z corresponds to Down). Helicopter linear velocities:v_N, v_E, v_D (North, East, Down). Helicopter linear locations: x_N, x_E, x_D (North, East, Down).

The helicopter is a multidimensional and non-linear plant. Therefore in the synthesis of the control system the knowledge of the main and coupling channels was applied [1], [2], [3], [4], [5].

The synthesis of the first control level uses the identification of the only selected main channels from input to output of the plant.

The idetification method based on a step response [7] was used herein.

Step responses for the selected channels from input to output of the plant are shown in Figures 2 to 4.

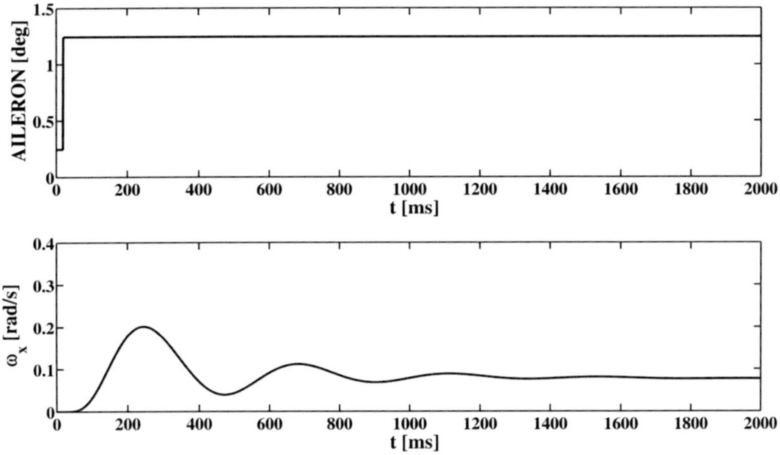

Fig. 3 The transient ω_x in the system excited by the unit step AILERON signal

Figure 2 presents a step response when the input signal is the step of AILERON signal and output signal is ω_x.

Figure 3 presents a step response when the input signal is the step of ELEVATOR signal and output signal is ω_y .

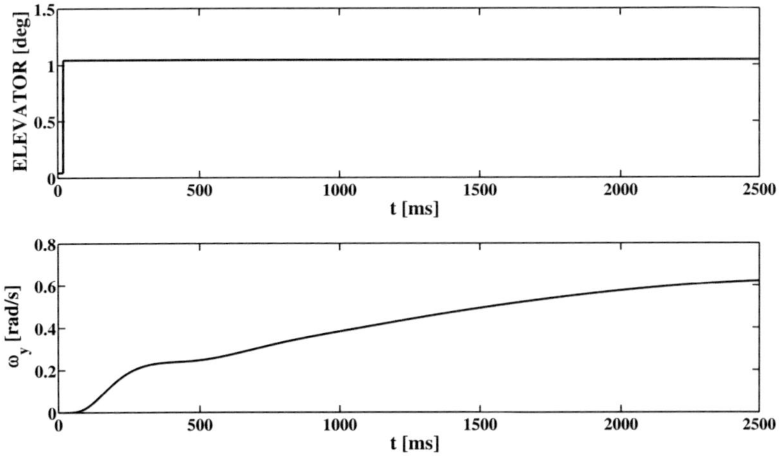

Fig. 4 The transient ω_y in the system excited by the unit step ELEVATOR signal

Figure 4 presents a step response when the input signal is the step of RUDDER signal and output signal is ω_z .

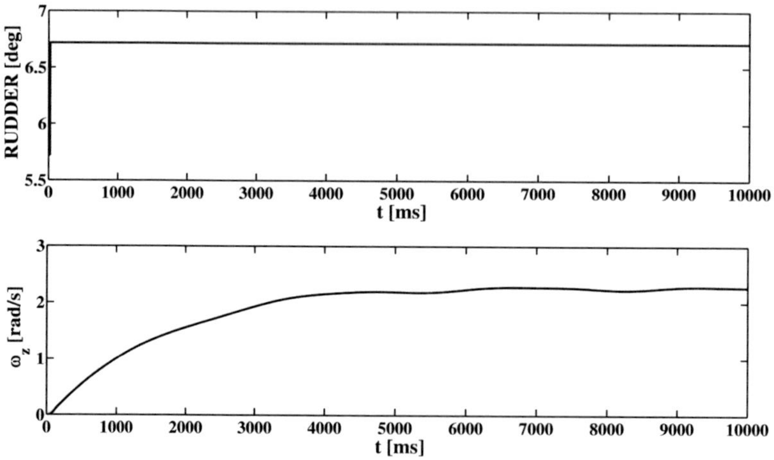

Fig. 5 The transient ω_z in the system excited by the unit step RUDDER signal

The next step is the identification of some transfer functions on the basis of the obtained step responses. These signals were used to identify transfer functions presented in (1), (2), (3) and (4). Hence we obtain transfer functions as follows:

$$K_{92}(s) = \frac{\omega_x(s)}{U_{AI}(s)} = \frac{1}{11.7 + 0.1672s + 0.05991s^2} \tag{1}$$

$$K_{103}(s) = \frac{\omega_y(s)}{U_{EL}(s)} = \frac{1 + 0.7641s}{0.1161 + 4.121s + 0.05758s^2 + 0.03567s^3} \tag{2}$$

$$K_{114}(s) = \frac{\omega_z(s)}{U_{RU}(s)} = \frac{1}{0.3427 + 0.8046s} \tag{3}$$

$$K_{45}(s) = \frac{U_{RU(s)}}{\omega_z(s)} \cdot \frac{\omega_z(s)}{U_{CO}(s)} = \frac{0.3427 + 0.9072s + 0.2408s^2}{0.3328 + 1.068s + 0.4260s^2} \tag{4}$$

The above transfer functions we obtain were obtained by means of the Matlab functions.

3 First Control Level

The first control level concentrates on angular velocity in three helicopter rotation axes.

This stage uses the identification of the selected main channels from input to output of the plant as presented above.

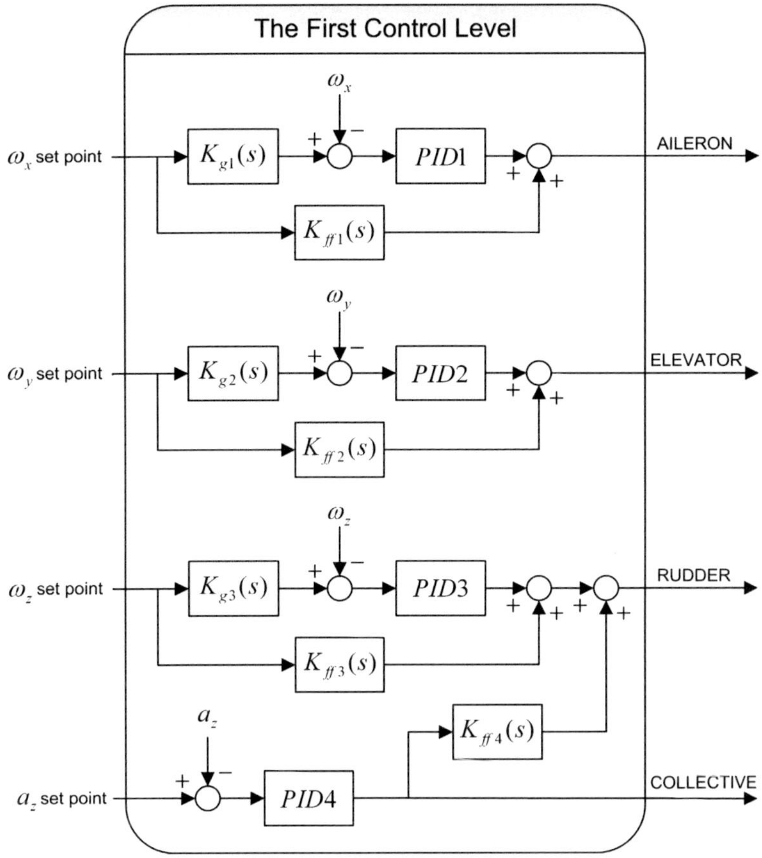

Fig. 6 The first control level

Moreover, the properties of some plant coupling channels, mentioned in the literature, are taken into account during the synthesis of control system [1], [2], [3], [4], [5].

Figure 5. presents first control level.

A feedback and a feedforward were used in the control system.

In Figure 6 a schematic diagram of PID controller is shown. Table 1 presents PID regulators settings. The regulators were tuned in a purely experimental way [6].

Table 1 PID settings

	PID1	PID2	PID3	PID4
P	2	3	9	0.5
I	2.5	30	25	10
D	0.2	0.1	0	0

Table 2 PID settings

	PID5	PID6	PID7
P	3.5	2.5	3
I	0	0	0
D	0	0	0

We use a first order lag as a filter transfer functions $K_g(s)$:

$$K_{g1}(s) = K_{g2}(s) = K_{g3}(s) = \frac{1}{(1+sT)},\ T = 0.1$$

These transfer functions allow to obtain control signals (AILERON, ELEVATOR, RUDDER, COLLECTIVE) without sudden changes.

The feedforward transfer functions $K_{ff}(s)$ were obtained as presented below:

$$K_{ff1} = \frac{K_{g1}(s)}{K_{92}(s)};\ K_{ff2} = \frac{K_{g2}(s)}{K_{103}(s)};\ K_{ff3} = \frac{K_{g3}(s)}{K_{114}(s)};\ K_{ff4} = K_{g4}(s) \cdot K_{45}(s);$$

4 Second Control Level

The second level focuses on angles of rotation control in three helicopter rotation axes.

This level is responsible for maintaining and setting the helicopter to the set orientation.

The helicopter attitude is described by three angles of rotation around X, Y, Z axes - Φ, Θ, Ψ.

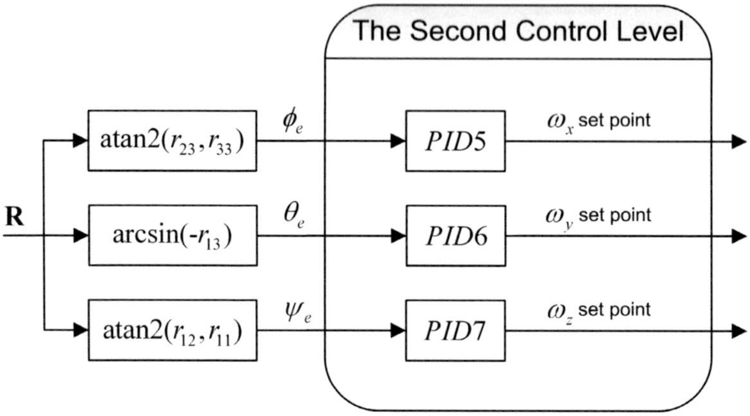

Fig. 7 The Second control level

Figure 7 presents a schematic diagram of the second control level. There is also shown the method of error angles Φ_e, Θ_e and Ψ_e calculation with using the attidude error matrix **R**.

Table 2 presents PID regulators settings. The regulators were tuned in a purely experimental way [6].

Helicopter attitude angles errors are calculated on the basis of the attitude error matrix as shown in Figure 7.

There is an additional block between second and third level which is not directly connected with control. The block transforms the data from the linear system to rotational system.

Details of the calculations are presented in Figures 7, 8 and 9.

Euler angles were also presented in a matrix form - $\mathbf{M_C}$ as shown in Figure 8. $\mathbf{M_C}$ defines current helicopter attitude. $\mathbf{M_D}$ defines set helicopter attitude.

Attitude error matrix **R** was used in this block. This matrix was built on the basis of set attitude matrix and Euler angles as was shown in Figure 8. The angles Φ, Θ and Ψ describe current helicopter attitude.

The block which was presented in Figure 9 also contains set attitude matrix . It was developed on the basis of helicopter linear set accelerations (**u**) and helicopter front set orientation (**f**). Helicopter linear set accelerations are determined in the coordinate system which is connected with the ground.

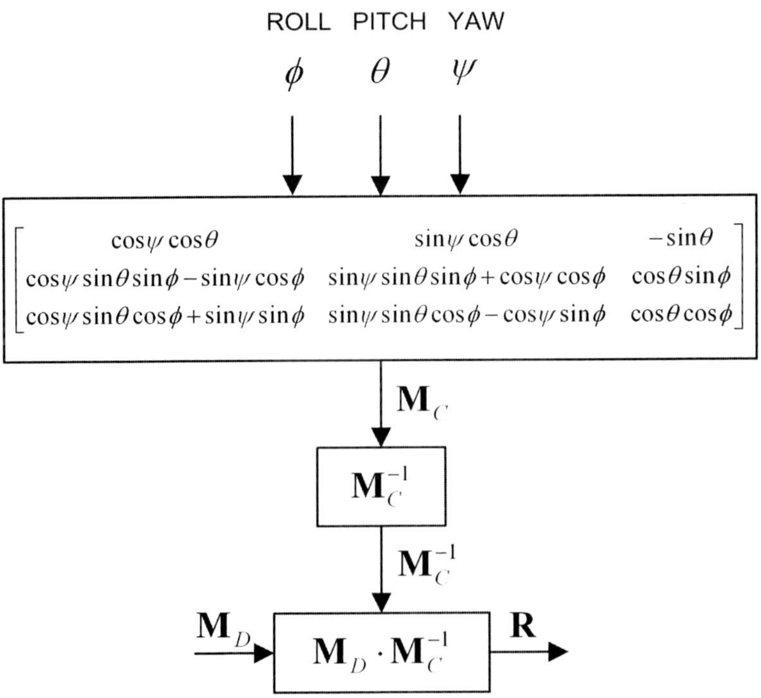

Fig. 8 The block of the orientation error matrix **R** calculation

Furtheron, attitude set matrix is calculated on the basis of the vectors **u** and **f**. These vectors determine desirable helicopter orientation. The top direction of the helicopter is defined by the vector u in the coordinate system which is connected with the ground. The vector **f** defines rotation around the *Z* axis in the coordinate system which is connected with the helicopter.

Gram-Schmidt process allows us to obtain the orthogonal vectors **u2** and **f2**. This process removes component in the specified direction without taking into account the vector length. The next step is to calculate a new vector (**f2** × **u2**) using the vector product. We obtain the unit vectors **u2**, **f2** and (**f2** × **u2**) which are result of normalization. These unit vectors describe new transformation, namely the helicopter attitude set matrix ($\mathbf{R} = [\mathbf{f2}; \mathbf{u2} \times \mathbf{f2}; \mathbf{u2}]$).

Due to normalisation we obtain the unit vectors **u2**, **f2** and (**f2** × **u2**).

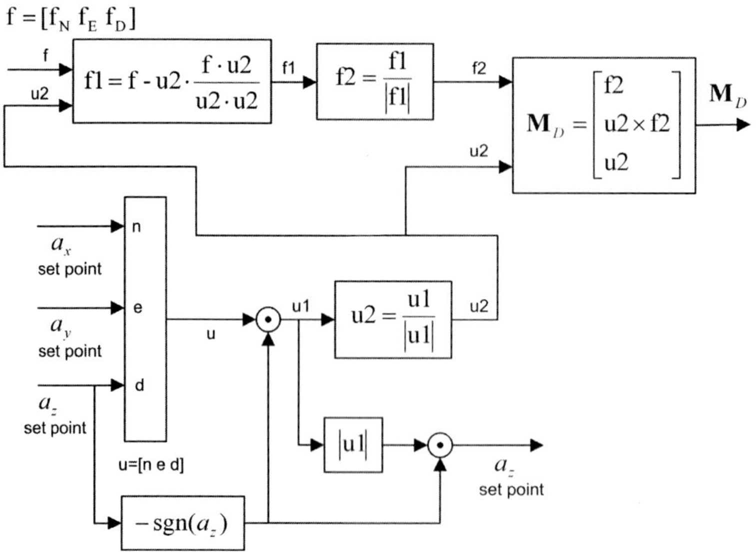

Fig. 9 The block of data transformation from the linear system to rotation system

5 Third and Fourth Control Levels

The third level of controllers calculates the set accelerations in the coordinate system which is connected with the ground. The calculations are made on the basis of the current and set helicopter velocities. All calculations are made in the same coordinate system.

The fourth level of controllers calculates the set velocities in the coordinate system which is connected with the ground. These calculations are made on the basis of the current and set helicopter locations.

Figure 10 shows a schematic diagram of the third and fourth control levels.

Table 3 presents PID regulators settings. The regulators were tuned in a purely experimental way [6].

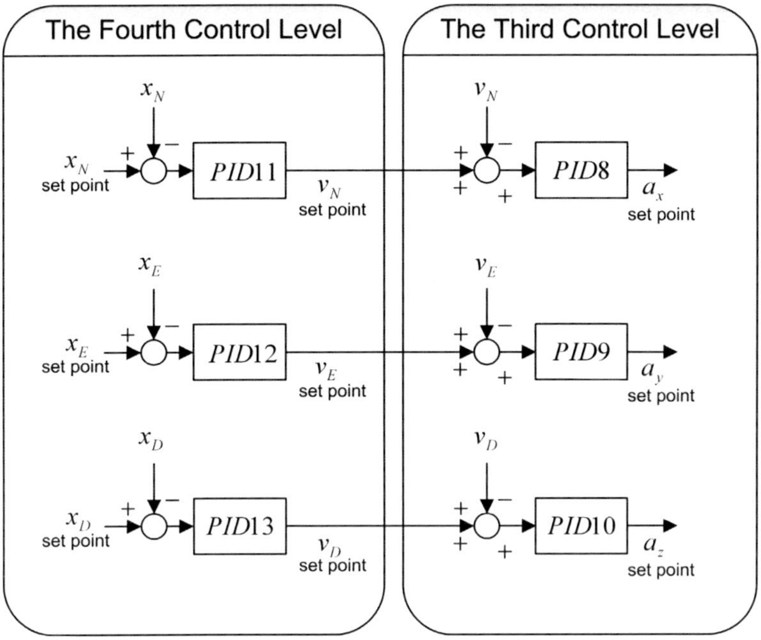

Fig. 10 Third and fourth control levels

Table 3 PID settings

	PID8	PID9	PID10	PID11	PID12	PID13
P	0.5	0.5	3	0.45	0.45	2
I	0	0	3	0	0	0
D	0	0	0	0	0	0.2

6 Examples of Helicopter Flight Symulations

Figures 11 to 13 demonstrate the example of the helicopter trajectory. The same trajectory is there shown in three axes - North-South, East-West and Down-Up.

Figure 11 presents the behavior of the helicopter during the flight. It can be observed that the current value of x_N, where x_N is the location of the helicopter along North axis, follows x_N set point.

Figure 12 presents the behavior of the helicopter during the flight. It is observed that current value of x_E, where x_E is the location of the helicopter along East axis, follows x_E set point.

Figure 13 presents the behavior of the helicopter during the flight. It is observed that current value of x_D, where x_D is the location of the helicopter along Down axis, follows x_D set point.

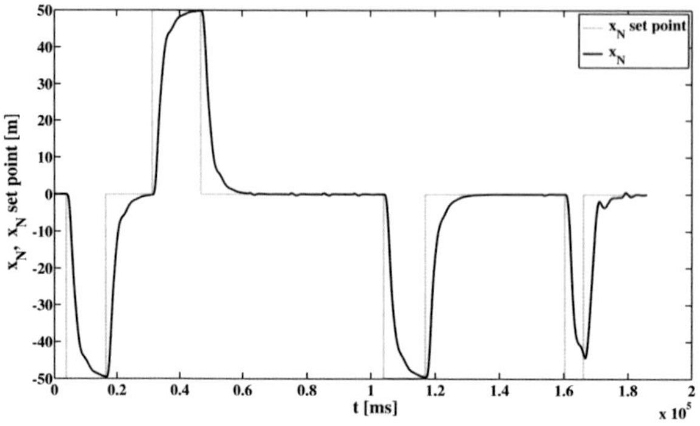

Fig. 11 The transient x_N set point and x_N during the helicopter flight

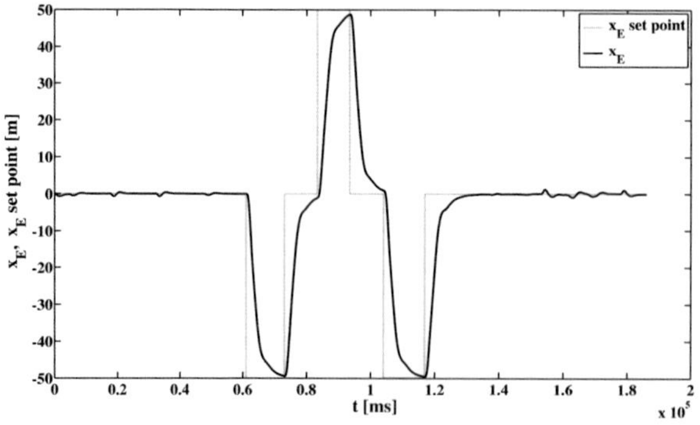

Fig. 12 The transient x_E set point and x_E during the helicopter flight

The example which was presented herein demonstrate a correct behavior of the flying helicopter model. It can be noticed that the step responses demonstrate that due to multilevel control system it is possible to achieve desired responses. In this case, the desired response to a step input is a first order response. Conclusively, it allows to achieve the 'smooth' trajectory of the flight.

Therefore the flight is safe. The small oscillations in the z axis direction, which can be observed in Figure 13, do not pose a threat for the helicopter.

The example which was presented herein demonstrates a correct behavior of the flying helicopter model.

The set trajectory consist of several step changes in the set point.

The possibility of introducing the set location in the coordinate system, which is connected with the ground, allows to reach the set geographical location by the helicopter in an auto mode.

The helicopter can track the set flight trajectory when the trajectory is built from the successive set geographical locations and the altitudes.

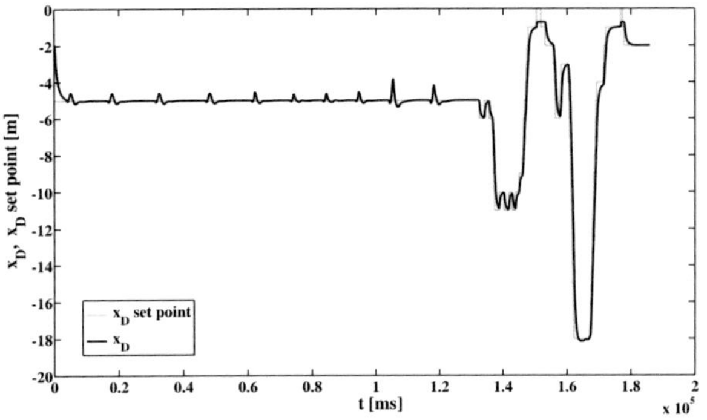

Fig. 13 The transient x_D set point and x_D during the helicopter flight

References

1. Nawrat, A.: Modelowanie i sterowanie bezzałogowych obiektów latajacych. Wydawnictwo Politechniki Śląskiej, Gliwice (2009)
2. Valavanis, K.P. (ed.): Advances in Unmanned Aerial Vehicles. Springer (2007)
3. Castillo, P., Lozano, R., Dzul, A.E.: Modelling and Control of Mini-Flying Machines. Springer (2005)
4. Padfield, G.D.: Helicopter Flight Dynamics. Backwell Science Ltd. (1996)
5. Manerowski, J.: Identyfikacja modeli dynamiki ruchu sterowanych obiektów latajacych. Wnaskon, Warszawa (1999)
6. Gessing, R.: Control Fundamentals. Wydawnictwo Politechniki Śląskiej, Gliwice (2004)
7. Kasprzyk, J. (ed.): Identyfikacja procesów. Wydawnictwo Politechniki Śląskiej, Gliwice (2002)

The Control System for Autonomous Mobile Platform

Artur Babiarz, Krzysztof Jaskot, and Przemysław Koralewicz

Abstract. Autonomous mobile platform is a machine that can operate in a human-made environment. The paper presents design of autonomous mobile platform based on the all terrain 1/8th scale four wheel drive radio control model. In this paper was considered problem of automatic control of mobile platform using information from GPS system, electronic compass and encoder. The mobile platform is equipped in two-stroke glow engine, heavy-duty drive train and wide-track suspension and controller based on ARM7 microcontroller and using MaxStream XBee Pro 2.4GHz radio modem communication module. The base station equipment is also described. Communication protocol between mobile platform and base station is presented. The paper presents also an application of electronic compass to measure azimuth of mobile platform. Problem of speed and distance control is described. Results of real application are also shown. Results of work on autonomous mobile platform that can operate in human environment are presented. The obtained properties of the system have been affected that it can be used for future research and autopilot design project.

Keywords: mobile platform, sensors, communication, microcontroller.

1 Introduction

The aim of the project was to create an autonomous mobile platform, which could operate in open terrain. Autonomous mobile platform is a machine that can operate

Artur Babiarz · Krzysztof Jaskot
Silesian University of Technology, Institute of Automatic Control,
Akademicka 16, 44-101 Gliwice, Poland
e-mail: {artur.babiarz,krzysztof.jaskot}@polsl.pl

Przemysław Koralewicz
ABB Research Center, Starowislna 13A, 31-038 Krakow, Poland
e-mail: przemyslaw.koralewicz@pl.abb.com

A. Nawrat et al. (Eds.): Advanced Technologies for Intelligent Systems, SCI 440, pp. 15–28.
springerlink.com

in a human-made environment [4, 5, 6, 7, 8, 11]. The key to autonomy is a control system built on the basis of information concerning the position and goal. Use GPS NMEA (National Marine Electronic Association) protocol allows you to obtain information in text form about the current location of the object. After adding information about the intermediate target points (Waypoint), we can receive information about the current direction [9, 12]. After the experiments conducted and described in the article [2] been amended accordingly to increase the accuracy of determining the direction of movement and speed control. The work was considered problem use information derived from the GPS and IMU (Inertial Measurement System) [3, 13] as a source of control signal. In addition, we uses also information from the magnetic rotation sensor - encoder. IMU currently provides only information from the electronic compass. This information is needed to determine the azimuth of mobile platform.

2 Mobile Platform

As described in the article [2] mobile platform has been built using a remote-controlled car on a scale 1/8th (length 55cm, width 43cm) and it was delivered by the HPI Racing. We selected terrain model with an independent suspension and four-wheel drive (4WD), because we wanted to create an autonomous platform that can operate in open terrain. The appearance of old version of mobile platform with installed controller and the GPS system is shown in figure 1. As the propulsion system used in this model, two-stroke internal combustion engine with a capacity of 3.5cm3 and 2HP (Horse Power). This allows the dispersal model to speed about 60km/h. In addition to the chassis and drive train in the composition of the platform includes two servos, which are responsible for controlling the throttle/brake and

Fig. 1 Old version of mobile platform

course. These two servos give us the ability to control the platform traction. After the experiments described in the article [2] we change the appearance of mobile platform. We add aluminum frame with installed GPS and 2.4GHz antenna. Aluminum frame prevents any interference generated by working servos, engine (gearbox). We also add new controller and IMU system. The appearance of new version of mobile platform is shown in figure 2.

Fig. 2 New version of mobile platform. 1- aluminum frame, 2- GPS, 3- antenna 2.4Ghz, 4- main controller, 5- IMU, 6- batteries

3 Main Controller

Figure 3 shows a block diagram of the main controller. The heart of the autonomous vehicle is mounted in the car - controller will be built using a microcontroller AT91SAM7S, managing servo, all sensors, power supply and wireless communication. To build a control system for mobile platform we uses microcontroller AT91SAM7S256 (ARM7 Core) it was delivered by the ATMEL. He is responsible for collecting information from sensors, processing them, the exercise of control algorithms and generate servo control signals. In addition, it enables wireless communication, reading values of the individual, shared variables and modify hem. Basic features of the microcontroller is: RISC architecture, maximum speed clock 60 MHz 256 KB FLASH program memory, 64KB SRAM memory, interrupt controller, three 16-bits timers, four 16-bits PWM modules, three USART interface,

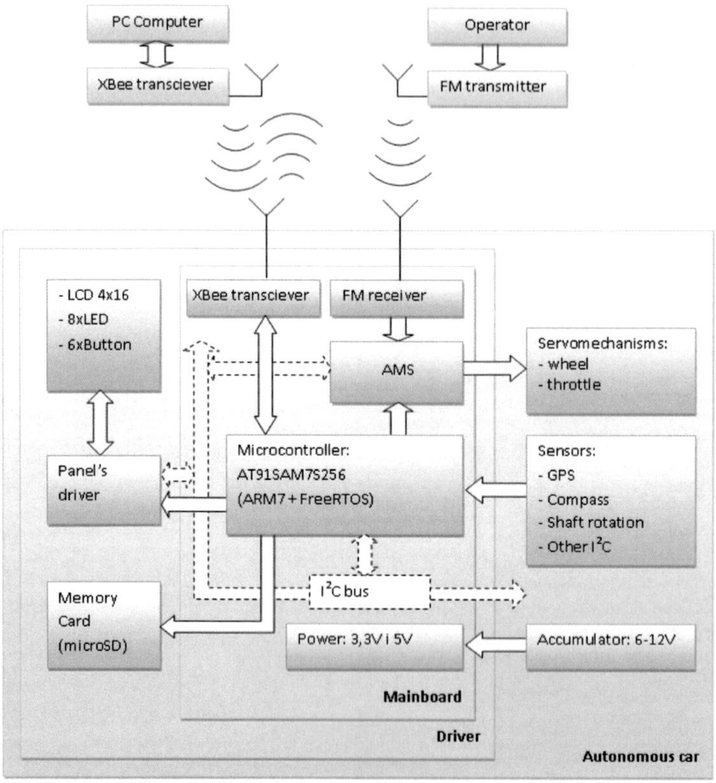

Fig. 3 Block diagram of the controller

USB controller, I2C bus, two SSP, SPI bus, 11-channels 10-bits A/D converter [14]. The microcontroller of our choice has also some other interesting features, such as on board USB controller that together with SAM-BA Boot Assistant provides an easy and fast way of programming the ARM. We have also the JTAG interface, which provides hardware debugging capabilities.

The main functions of the controller are: collecting information from sensors, generating servo control signals through the implementation of control algorithms based on readings from sensors, providing full duplex wireless communication with a PC (telemetry, simulation), I2C management. The process of autonomous control based on sensor readings of the following:

1. GPS - provides information on the location of the car on the globe, can also be a source of information about vehicle speed and orientation relative to the direction north.
2. Electronic compass (IMU) - allows you to specify the orientation of the car towards the north.
3. Rotation sensor - mounted on the drive shaft provides information about speed of the car, used as a source of feedback for the speed.

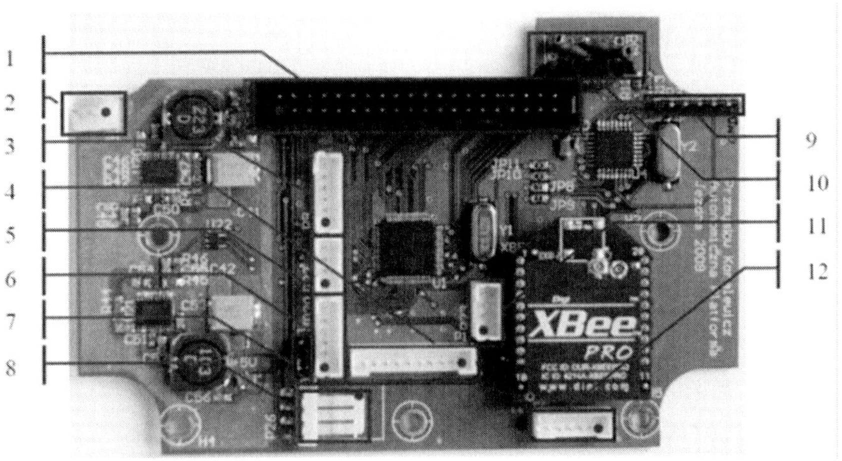

Fig. 4 Main controller

Other sensors, such as: temperature sensors [1], ultrasonic rangefinders, may be connected to an external derived I2C bus driver. Main controller using the AT91 microcontroller is shown in figure 4.

Where: 1 - operator panel connector, 2 - power connector, 3 -UART2 connector, 4 - JTAG connector, 5 - I2C connector, 6 - UART0 connector, 7 - UART0 power jumper, 8 - servo connector, 9 - ISP connector, 10 - radio receiver connector, 11 - rotation sensor connector, 12 - UART1 connector.

The controller is equipped with a plate operator panel with LCD, LEDs and push buttons for menu operation serving (figure 5) where: a) top view b) bottom view, 1 - operator panel connector, 2 - LCD HD44780 connector, 3 - place for display screen, 4 - S1-S6 - buttons, 5 - D1-D8 - LEDs, 6 - LEDs indicate the presence of voltage, 7 - reset button AT91SAM7S, 8 - LED connected to AT91SAM7S, 9 - MicroSD slot, 10 - MicroUSB, 11 - JTAG, 12 - ISP.

With them you can read some of the performance of the program, namely: as GPS data, compass reading. During normal operation, the buttons are not available because the case is sealed. Then the display shows the next cycle menus that contain the most relevant data. Sample screens are shown in figure 6.

3.1 Autonomous/Manual Switch

The aim of the work is to create an autonomous vehicle, but at the prototype stage and testing is necessary to reduce its autonomy and ensure operator efficiency even take control of the car, which for various reasons, may start to behave unpredictably. Application of five-channel FM RC controller allowed the design of switch mode AMS (Autonomous/Manual Switch).

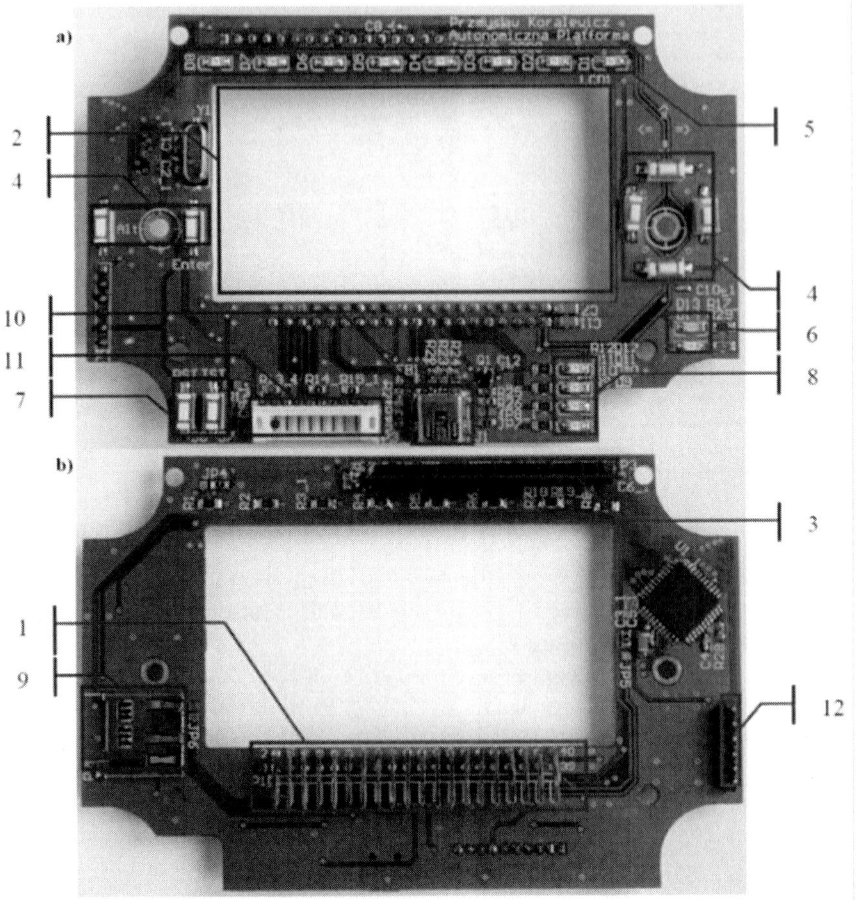

Fig. 5 Operator panel

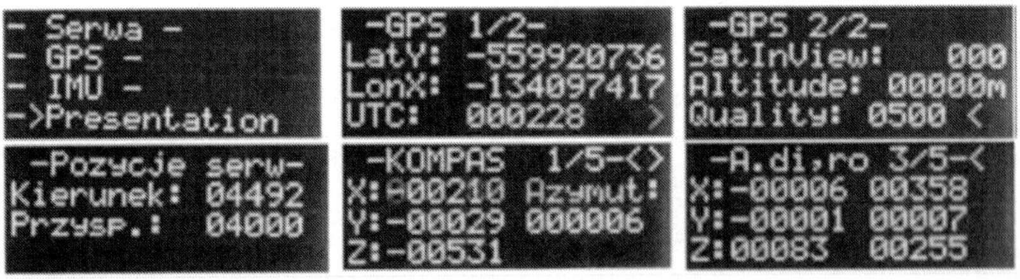

Fig. 6 User interface (from left: main menu, GPS 1/2, GPS 2/2, servo position, electronic compass, accelerometer)

Two channels, as in the case of a simple remote control, are used to transmit information about the location and direction of the throttle servo. A third channel, acts as a switch to allow the change of source signals servomechanisms between the

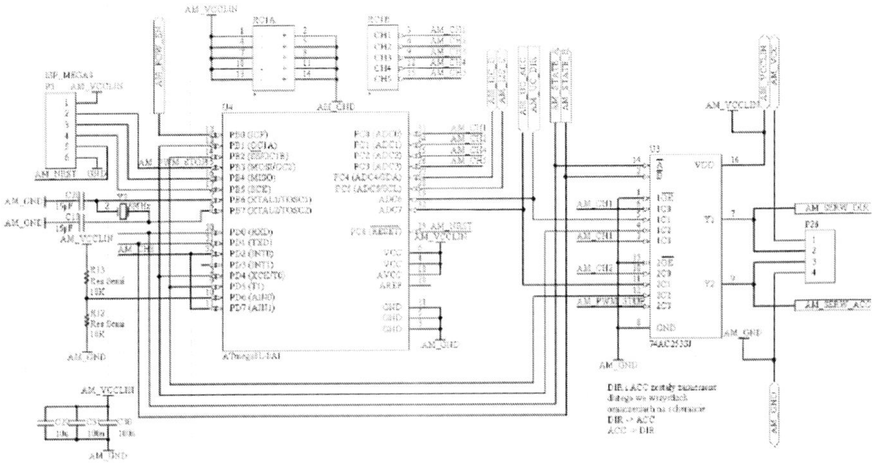

Fig. 7 Auto/Manual switch

Fig. 8 Connectors view. 1 - power connector, 2 - UART0 connector (IMU), 3 - UART2 connector (GPS), 4 - rotation sensor and I2C, 5 - servo connector, 6 - antenna connector 2,4GHz

apparatus of the FM (manual) and microcontroller (autonomous). In addition, upon failure of the FM signal from the apparatus or GPS, followed by AMS emergency stop of the vehicle. AMS switch is built using 8-bit microcontroller from Atmel-ATMega8. It is also connected to the I2C bus of main controller. The electrical diagram of AMS is shown in figure 7. The controller consists of two mounted in a housing designed for the project PC Band LCD. In total, this creates a three-layer structure. Connection with external devices is via connectors placed on the body (figure 8).

Furthermore, a battery of dimensions 65x40x10mm can be mounted inside the controller. Figure 9 shows the main controller mounted on the mobile platform.

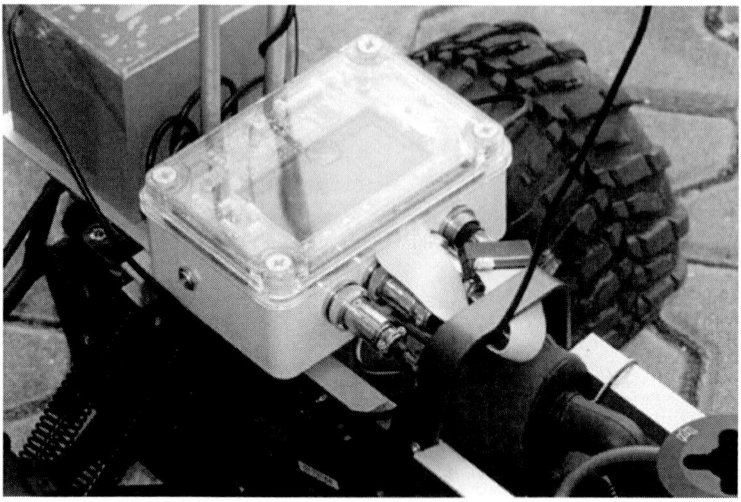

Fig. 9 Main controller mounted on a mobile platform

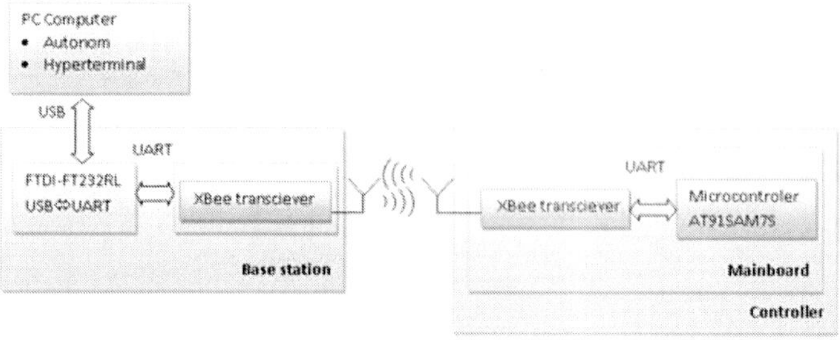

Fig. 10 Full duplex communication between PC and controller

3.2 Communications

In addition to the equipment fitted on mobile platform available to the operator is base station, which includes: remote control transmitter - manual control and auto/manual switch, radio communication module XBee PRO 2.4GHz - sending/receiving data to/from the vehicle, computer witch control application providing remote viewing operating parameters of the control system mounted on a platform. Figure 10 shows a block diagram system using wireless communication modules, XBee PRO. These modules operate in transparent mode. This means that any data sent to the first module will be unchanged at the output of the second module. Communication with the modules is done using a UART protocol, which is easy to implemented in the microcontroller AT91SAM7S. From the PC side it is necessary to convert the signals. After the application of FT232RLFTDI and appropriate drivers on your computer, you can create a virtual serial port.

Fig. 11 Example of a data frames - request

Fig. 12 Example of a data frames - response

Communication between the controller and the computer is realized in the form of data frames. An example of a data frame containing the request and response are shown in figures 11, 12. This frame allows you to read off the value of n variables. In the body frame includes a request ID variable in hexadecimal, separated by commas. The answer of controller is frame containing a comma-separated variables. During operation of an autonomous platform, it is possible to preview and recording parameters on the PC using communication modules.

4 Control System

ARM7TDMI processor architecture allows you to install an RTOS (Real Time Operating System). The operating system allows the division of tasks performed by the microcontroller processes. All tasks are performed simultaneously which is referred to as multitasking. The operating system also provides mechanisms for communication between processes and synchronization tasks. The core of the system takes over responsibility for the allocation of CPU time for the process, taking into account the priorities of processes and supporting a notification of interruption.

The application of controller operates under the FreeRTOS system [15]. It is optimized for embedded systems with low hardware resources consists of only three source files written in C language. FreeRTOS also makes dynamic memory management, create lists and introduces features to process execution pauses for a specified period of time. Figure 13 shows the diagram of mechanisms for exchange of data between the key elements of the microcontroller software to which they are: processes-constantly running in a loop programs, drivers-used, among others by processes to communicate with microcontroller peripherals. Autonomous vehicle program consists of 6 processes which have been divided between tasks. The process of "GPS" is responsible for operating the GPS receiver. When you receive the correct line of NMEA0183 (every 5ms) is processed, and read data update the corresponding variables. The process of "IMU" collects data from the IMU device driver via UART0, every 20ms is sent to query the value of the magnetic compass reading. The process of "Interface" is used to refresh the driver of the operating panel is currently displayed on the LCD. The process of "LED" acts as a signal generator, reporting via LEDs on the correct behavior. The process of "Control" uses data to

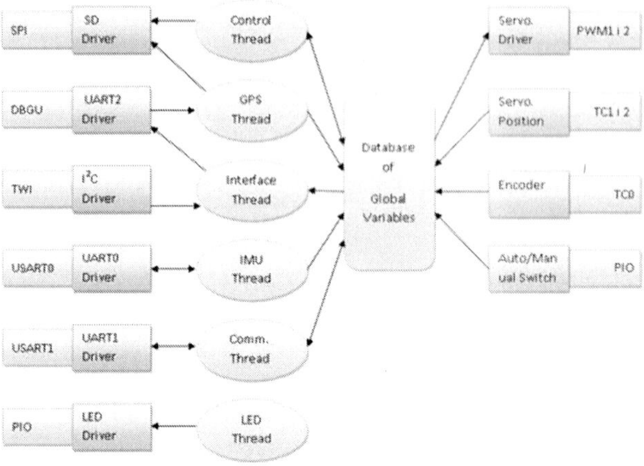

Fig. 13 Division of microcontroller programs and threads used for peripherals and their drivers

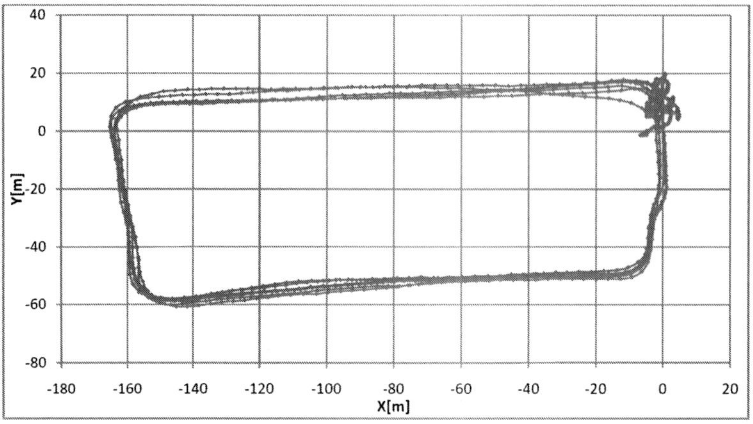

Fig. 14 Results of test GPS system

assess the state in which currently there is a car (including position, speed, azimuth), and then using the implemented algorithms update a servo settings. The process of "Communication" is the basis of telemetric system of the car.

5 Results

Test drives using the built controller were implemented in several stages. The first was the verification of the system of GPS and radio communications transmission measurements to the base station. At this stage, was carried out manual control of

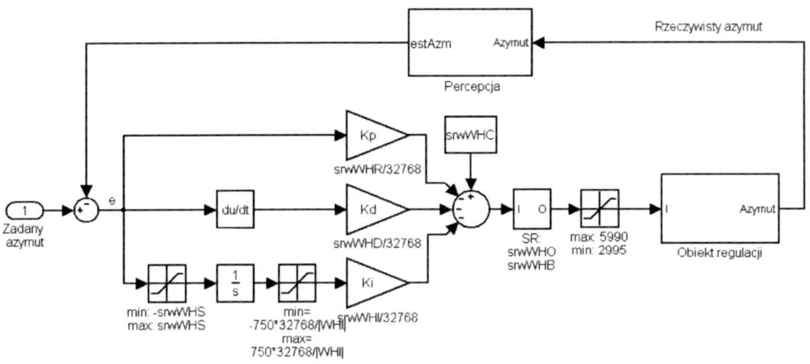

Fig. 15 Block diagram of the azimuth control system

direction and throttle. The results of this test are shown in figure 14. Test results for the manual control shows a high accuracy to obtain information from GPS. The second test phase was to incorporate automatic direction and throttle control. Information on the target points WP1 - WP7 was recorded in the memory of the microcontroller. In an exemplary control algorithm using two independent controllers. First, the block diagram shown in figure 15, controls the servo direction. It seeks to minimize the difference between the azimuth obtained in the calculations in the perception stage, and a given azimuth. The latter is the direction from which the car should move to reach the waypoint in a straight line. This system is a PID regulator, extended with additional capabilities.

The second controller shown in figure 16 controls the speed servo uses a PI regulator. Preset speed depends on the quality of data obtained in the stage of perception. It is estimated based on the number of satellites used by the GPS receiver to measure the position. If their number is less than four, the quality is considered satisfactory and speed reference is 0. Otherwise, it is different from zero and can take two values. The value of fast (1m/s) applies only when the car is already headed toward the via point and not located closer than 2m from it. If the distance is less than 2m predetermined value of speed is 0.25m/s. This helps the controller to turn the direction of the correct execution.

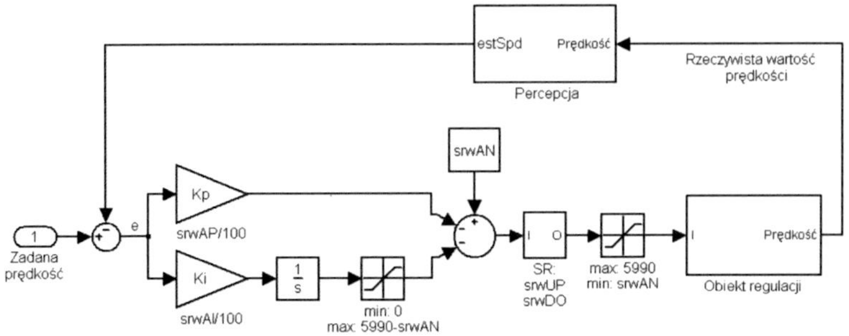

Fig. 16 Block diagram of speed control

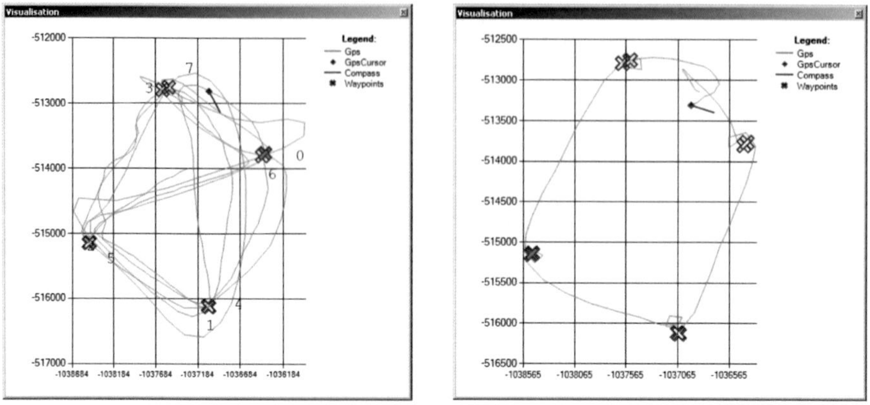

Fig. 17 Examples of waypoints navigation

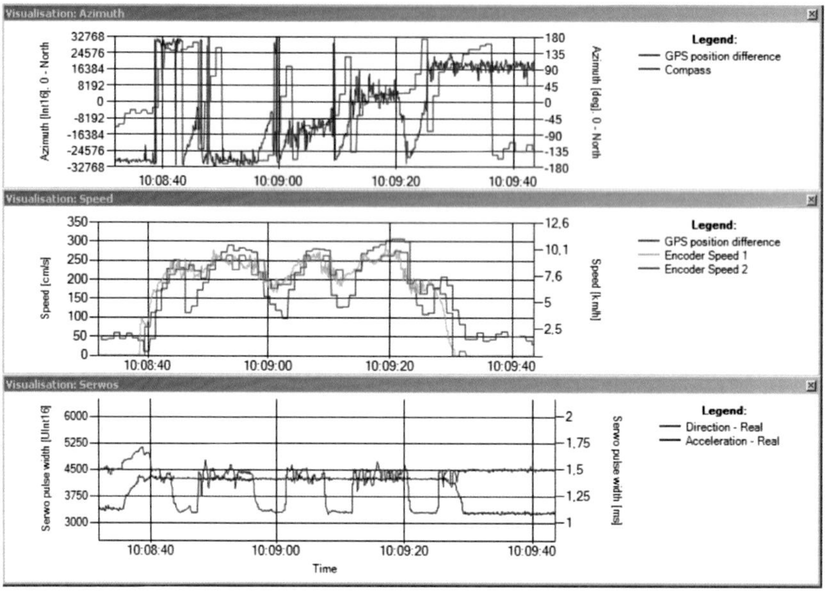

Fig. 18 Examplary measurments recorded during autonomous drive (Azimuth -GPS, electronic compass; Speed - GPS, encoder; Servos)

Examples of autonomous waypoints navigation (figure 17) show positions and orientation taken from GPS and electronic compass. During drive we also recorded measurements such as velocity (GPS, encoder), azimuth (GPS, electronic compass), servos (figure 18).

In order to protect the engine from overheating we used 18B20 temperature sensor. It communicates with the microcontroller using 1-wire bus, measurements were

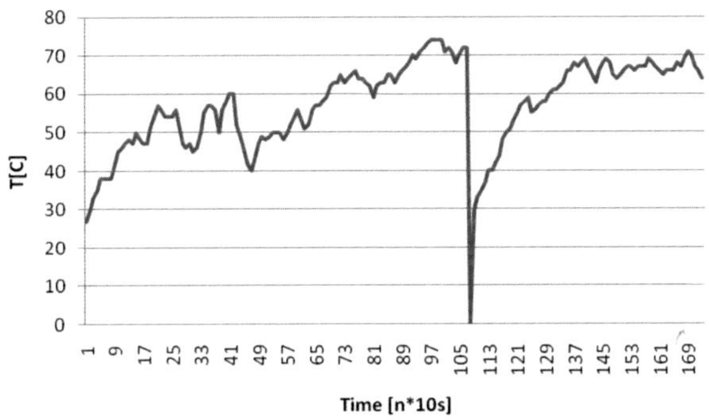

Fig. 19 Temperature of engine

made every 10 seconds [1]. Figure 19 shows the examplary temperature recorded during the driving mobile platform.

6 Summary

The proposed solution allows to control an autonomous mobile platform based on the all terrain 1/8th scale four wheel drive radio control model using a GPS, electronic compass and encoder. All control algorithms were implemented by the main controller built using a microcontroller with an ARM7 core and worked under the control of FreeRTOS. Currently, the platform does not have the possibility of overtaking obstacles. Further development of the mobile platform will be implemented will support the addition of sensors (e.g. laser scanner, ultrasonic rangefinder), allowing to overcome obstacles and to develop new control algorithms [10]. After applying the sensor system may need to change the glow engine in the electric motor with reverse gear.

References

1. Babiarz, A., Jaskot, K.: Temperature control system for glow engine. In: Zeszyty Naukowe Politechniki Śląskiej. Automatykaz, Gliwice, vol. 150, pp. 115–121 (2008)
2. Babiarz, A., Jaskot, K.: Autonomous Mobile Platform. In: International Carpathian Control Conference, ICCC 2010, Eger, Hungary, pp. 179–183 (2010)
3. Babiarz, A., Jaskot, K.: The inertial measurement unit for detection of position. Electrical Review R. 86(11a), 323–334 (2010)
4. Baker, C.R., Dolan, J.M.: Street Smarts for Boss - Behavioral Subsystem Engineering for the Urban Challenge. IEEE Robotics and Automation Magazine (2009)
5. Behringer, R., Maurer, M.: Results on Visual Road Recognition for Road Vehicle Guidance. In: Proceedings of the 1996 IEEE Intelligent Vehicles Symposium, Tokyo, Japan (1996)

6. Braun, T., Schd'żf́er, H., Berns, K.: Topological Large-Scale Off-road Navigation and Exploration - RAVON at the European Land Robot Trial 2008. In: The 2009 IEEE/RSJ International Conference on Intelligent Robots and Systems, St. Louis, USA (2009)
7. Freire, E., Bastos-Filho, T., Sarcinelli-Filho, M., Carelli, R.: A new mobile robot control approach via fusion of control signals. Part B: IEEE Transactions on Cybernetics Systems, Man, and Cybernetics 34(1), 419–429 (2004)
8. Harkins, R., et al.: Design and testing of an autonomous highly mobile robot in beach environment. In: Proceedings of the World Congress on Engineering and Computer Science 2008, WCECS 2008, San Francisco, USA (2008)
9. Jaskot, K.: Implementation of GPS information to control of UAV model. ZN Politechniki Rzeszowskiej, Mechanika, Awionika, Rzeszów, vol. 71 (2007)
10. Kuwata, Y., Karaman, S., Teo, J., Frazzoli, E., How, J.P., Fiore, G.: Real-Time Motion Planning With Applications to Autonomous Urban Driving. IEEE Transactions on Control Systems Technology 17(5), 1105–1118 (2009)
11. Li, T.-H.S., Chang, S.-J., Tong, W.: Fuzzy target tracking control of autonomous mobile robots by using infrared sensors. IEEE Transactions on Fuzzy Systems 12(4), 491–501 (2004)
12. Shair, S., Chandler, J.H., Gonzalez-Villela, V.J., Parkin, R.M., Jackson, M.R.: The Use of Aerial Images and GPS for Mobile Robot Waypoint Navigation. IEEE/ASME Transactions on Mechatronics 13(6), 692–699 (2008)
13. Soloviev, A.: Tight Coupling of GPS and INS for Urban Navigation. IEEE Transactions on Aerospace and Electronic Systems 46(4), 1731–1746 (2010)
14. AT91 ARM Thumb-based Microcontrollers, `www.atmel.com`
15. FreeRTOS, ARM7, `http://www.freertos.org`

Lyapunov Exponents for Discrete Time-Varying Systems

Adam Czornik, Aleksander Nawrat, and Michał Niezabitowski

Abstract. The purpose of this paper is to present some results on the effects of parametric perturbations on the Lyapunov exponents of discrete time-varying linear systems. We fix our attention on the greatest and smallest exponents.

Keywords: difference equation, stability, time-varying systems, Lyapunov exponents, linearization.

1 Introduction

In the laast decade there has been great interest of researchers from the theory of linear models in systems which combine logical switches and differential or difference equations. This interest is dictated first and foremost of the great utility of such models for modeling real-world objects. The usefulness of this growing demand depends on the methods of modeling, analysis and understanding of this structure. Although construction of a hybrid model is a relatively simple task, its analysis is already far from simplicity. During the analysis of hybrid systems many interesting and difficult mathematical problems arise. Many of them are associated with the dynamics, and in particular the stability of such models remains unsolved today.

Many properties of dynamic systems can be successfully characterized by certain numbers called numerical characteristics or characteristic exponents. These include: Lyapunov, Bohl, Perron, Izobov, Grobman exponents, generalized spectral radiuses. These numbers describe the different types of stability, growth of trajectories or sensitivity on parametric disturbances.

Adam Czornik · Aleksander Nawrat · Michał Niezabitowski
Silesian University of Technology, Institute of Automatic Control,
Akademicka 16, 44-101 Gliwice, Poland
e-mail: {Adam.Czornik,Aleksander.Nawrat}@polsl.pl,
Michal.Niezabitowski@polsl.pl

A. Nawrat et al. (Eds.): Advanced Technologies for Intelligent Systems, SCI 440, pp. 29–44.
springerlink.com

Since the works of A. M. Lyapunov [40] and O. Perron [49, 50] the theory of Lyapunov exponents became the subject of intense research, as evidenced by the huge number of papers published on this subject.

This paper is devoted to the influence of parametric perturbations on the Lyapunov exponents of discrete linear systems with time varying coefficients. Different types of perturbations are considered, limited in terms of certain norms, tending to zero in a specified rate. For each of them we describe their impact on the value of Lyapunov exponents.

There are several monographs devoted in whole or in large part to the Lyapunov exponents, e.g. [4, 6, 12, 32, 41, 46]. But only in the last two a problem of parametric perturbations is discussed. Both of these items deals with continuous time systems, and in addition, they are available only in Russian.

2 Definitions of Characteristic Exponents and Basic Properties

In this section we will introduce the notation and basic properties of Lyapunov exponents of linear discrete time-varying systems:

$$x(n+1) = A(n)x(n), n \geq 0, \tag{1}$$

where $(A(n))_{n\in\mathbb{N}}$ is a bounded sequence invertible of s-by-s real matrices such that $\left(A^{-1}(n)\right)_{n\in\mathbb{N}}$ is bounded.

The transition matrix of (1) is defined as

$$\mathscr{A}(m,k) = A(m-1)...A(k)$$

for $m > k$ and $\mathscr{A}(m,m) = I$, where I is the identity matrix. For a initial condition $x(0) = x_0 \in R^s$ the solution of (1) is denoted by $x(n,x_0)$, so

$$x(n,x_0) = \mathscr{A}(n,0)x_0.$$

Definition 1. Let $b = (b(n))_{n\in\mathbb{N}}$ be a sequence of real numbers. The number (or the symbol $\pm\infty$) defined as

$$\lambda(b) = \limsup_{n\to\infty} \frac{1}{n} \ln|b(n)|$$

is called the upper characteristic exponent or simply characteristic exponent of sequence $(b(n))_{n\in\mathbb{N}}$. For a sequence $v = (v(n))_{n\in\mathbb{N}}$ of vectors of normed space $(X, \|*\|)$ we define its characteristic exponent $\lambda(v)$ as a exponent of sequence $(\|v(n)\|)_{n\in\mathbb{N}}$.

It is easy to check that finite $\lambda(b)$ is a characteristic exponent of sequence of $b = (b(n))_{n\in\mathbb{N}}$ if, and only if, the following two conditions are simultaneously satisfied:

1. For any $\varepsilon > 0$ there exists constant D_ε such that for all $n \in \mathbb{N}$ the following inequality is satisfied

$$|b(n)| \leq D_\varepsilon \exp((\lambda(b)+\varepsilon)n);$$

2. For any $\varepsilon > 0$ the following equality is satisfied

$$\limsup_{n\to\infty} |b(n)| \exp((-\lambda(b)+\varepsilon)n) = \infty.$$

Fix an arbitrary norm $\|\cdot\|$ in R^s and denote induced operator norm by the same symbol. Denote

$$\sup_{n\in\mathbb{N}} \|A(n)\| = a, \quad \sup_{n\in\mathbb{N}} \left\|A^{-1}(n)\right\| = a'. \tag{2}$$

Definition 2. For $x_0 \in R^s$, $x_0 \neq 0$ the Lyapunov exponent $\lambda_A(x_0)$ of (1) is defined as characteristic exponent of $(x(n,x_0))_{n\in\mathbb{N}}$ that is

$$\lambda_A(x_0) = \limsup_{n\to\infty} \frac{1}{n} \ln \|x(n,x_0)\|.$$

We also define $\lambda_A(0) = -\infty$.

Observe that by the equivalence of all norms in R^s the above definition does not depend on the particular choice of the norm. The next Theorem contains some basic properties of Lyapunov exponents.

Theorem 1. *For the Lyapunov exponents of (1) the following properties hold:*

1. *if $x_0 \in R^s$ and $c \in R$, $c \neq 0$ then $\lambda_A(x_0) = \lambda_A(cx_0)$;*
2. *if $x_1, x_2 \in R^s$ then $\lambda_A(x_1+x_2) \leq \max\{\lambda_A(x_1), \lambda_A(x_2)\}$;*
3. *if $x_1, x_2 \in R^s$ and $\lambda_A(x_1) \neq \lambda_A(x_2)$ then*

$$\lambda_A(x_1+x_2) = \max\{\lambda_A(x_1), \lambda_A(x_2)\};$$

4. *if $x_1, \ldots, x_l \in R^s \setminus \{0\}$ and the numbers $\lambda_A(x_1), \ldots, \lambda_A(x_l)$ are distinct, then the vectors $x_1, \ldots, x_l$ are linearly independent;*
5. *if $x_1, \ldots, x_s$ is a basis or R^s then*

$$\limsup_{n\to\infty} \frac{1}{n} \ln |\det \mathscr{A}(n,0)| \leq \sum_{l=1}^{s} \lambda_A(x_l); \tag{3}$$

6. *if $x_0 \in R^s$ then $\lambda_A(x_0) \leq a$;*
7. *if $x_0 \in R^s$ then $\lambda_A(s) \leq \lambda_A(x_0)$, where $v = (v(n))_{n\in\mathbb{N}}$ is given by*

$$v(n) = \begin{cases} \sum_{l=0}^{n-1} x(l,x_0) \text{ if } \lambda(x_0) \geq 0 \\ \sum_{l=n}^{\infty} x(l,x_0) \text{ if } \lambda(x_0) < 0 \end{cases}.$$

The proof of points 1-4 is given in [6], Theorem 2.1, inequality (3), which is called Lyapunov inequality was shown in [22], point 6 is a obvious consequence of (2) and the definition of $\lambda_A(x_0)$, point 7 is proved in [16], Lemma 4. As a consequence of point 4 we see that the set $\{\lambda_A(x_0) : x_0 \in R^s \setminus \{0\}\}$ contains at most s elements, say $-\infty \leq \lambda_1(A) < \lambda_2(A) < ... < \lambda_r(A) < \infty$, and the set $\{\lambda_1, \lambda_2, ..., \lambda_r\}$ will be called the spectrum of (1). The greatest and the smallest exponent we will denote by $\lambda_g(A)$ and $\lambda_s(A)$, respectively. An immediate consequence of definition of operator norm is the following result, which express the greatest exponent just in terms of the matrices $(A(n))_{n\in\mathbb{N}}$.

Theorem 2. *The greatest exponent $\lambda_r(A)$ of (1) is given by the following formula*

$$\lambda_g(A) = \limsup_{n\to\infty} \frac{1}{n} \ln \|A(n-1)...A(0)\|. \tag{4}$$

Together with (1) we will consider the so-called dual or adjoint system

$$y(n+1) = B(n)y(n), n \geq 0, \tag{5}$$

where $B(n) = \left(A^T(n)\right)^{-1}$. The transition matrix of the dual system is given by

$$\mathscr{B}(m,k) = B(m-1)...B(k)$$

for $m > k$ and $\mathscr{B}(m,m) = I$.

Observe that an application of Lyapunov inequality (3) to the dual system leads to the following extension of point 6 of Theorem 1.

Lemma 1. *If $x_0 \in R^s \setminus \{0\}$, then $\lambda(x_0)$ is finite.*

Further extension of this result can be found in [37]. It demonstrates the following conditions weaker than (2) conditions

$$\limsup_{n\to\infty} \frac{1}{n} \ln \|\mathscr{A}(n,0)\| < \infty$$

and

$$\limsup_{n\to\infty} \frac{1}{n} \ln \|\mathscr{B}(n,0)\| < \infty$$

imply the finiteness of $\lambda(x_0)$ for $x_0 \in R^s \setminus \{0\}$.

Denote by $\{\mu_1, \mu_2, ..., \mu_r\}$, $\mu_r < \mu_{r-1} < ... < \mu_1$ the spectrum of dual system. For each λ_i and μ_i we consider the following subspaces of R^s

$$E_i = \{v \in R^s : \lambda(v) \leq \lambda_i\}$$

and

$$F_i = \{v \in R^s : \mu(v) \leq \mu_i\},$$

and we set $E_0 = F_{s+1} = \{0\}$. The multiplicities n_i and m_i of Lyapunov exponent λ_i and μ_i are defined as $\dim E_i - \dim E_{i-1}$ and $\dim F_i - \dim F_{i+1}$, respectively for

$i = 1, \ldots, s$. If we have two bases $v_1, \ldots, v_s$ and $w_1, \ldots, w_s$ of R^s, then we will call them dual if $\langle v_i, w_j \rangle = \delta_{ij}$, where $< u, v >$ is the standard scalar product in R^s and δ_{ij} is the Kronecker symbol. For a base $V = \{v_1, \ldots, v_s\}$ of R^s we define the sum σ_V of Lyapunov exponents

$$\sigma_V = \sum_{i=1}^{s} \lambda_A(v_i).$$

The base $v_1, \ldots, v_s$ is called normal if for each $i = 1, \ldots, s$ there exists a basis of E_i composed of vectors $\{v_1, \ldots, v_s\}$. Formally, we should say that a basis is normal with respect to family E_i, $i = 1, \ldots, s$. It can be shown (see [7], remark after Theorem 1.2.5) that there always exist normal bases $v_1, \ldots, v_s$ and $w_1, \ldots, w_s$ (respectively of the families E_i and F_i) which are dual. It can be also shown (see [7], Theorem 1.2.3) that for the normal bases the sum σ_V of Lyapunov exponents is minimal and then, according to Lyapunov inequality (3), equal to

$$\limsup_{n \to \infty} \frac{1}{n} \ln |\det \mathscr{A}(n, 0)|.$$

For a basis $v_1, \ldots, v_s$ matrix $\mathscr{V}(n)$, $n \in N$ whose columns are $x(n, v_1), \ldots, x(n, v_s)$ is called fundamental matrix of (1). For a fundamental matrix the kernel $\mathscr{G}(n, m) = \mathscr{V}(n)\mathscr{V}^{-1}(m)$, $n, m \in N$ is called a Green's matrix of (1). If the base is normal, then the fundamental and Green's matrices are called normal. In many our further consideration a crucial role will be played by the possibility of reduction of our system to an upper triangular one. It is guaranteed by the following theorem from [7], Theorem 7.

Theorem 3. *For each sequence $(A(n))_{n \in \mathbb{N}}$ there exists a sequence $(U(n))_{n \in \mathbb{N}}$ of orthogonal matrices such that $C_n = U_{n+1}^T A_n U_n$ is upper triangular.*

Together with (1) we consider the following perturbed system

$$z(n+1) = (A(n) + \Delta(n))z(n), \tag{6}$$

where $\Delta = (\Delta(n))_{n \in \mathbb{N}}$ is a sequence of s-by-s real matrices from a certain class $\mathfrak{M}$. Under the influence of the perturbation Δ, the characteristic exponents of (1) vary, in general, discontinuously. It is possible that a finite shift of the characteristic exponents of the original system (1) corresponds to an arbitrarily small $\sup \|\Delta(n)\|$. In particular, it is possible for an exponentially stable system to be perturbed by an exponentially decreasing perturbation and the resulting system is not stable. The quantities

$$\Lambda_u(\mathfrak{M}) = \sup\{\lambda_g(A + \Delta) : \Delta \in \mathfrak{M}\}$$
$$\Lambda_l(\mathfrak{M}) = \inf\{\lambda_g(A + \Delta) : \Delta \in \mathfrak{M}\}$$

are referred to as the maximal upper and minimal lower movability boundary of the higher exponent of (1) with perturbation in the class $\mathfrak{M}$. We may also consider similar quantities for the others elements of the spectrum.

The determination of the movability boundaries of the higher exponent under various perturbations is one of the main problem in the theory of Lyapunov exponents. This problem has been studied for continuous-time systems for many classes $\mathfrak{M}$. For example, upper bound for the higher exponent of (1) under small perturbations, the so-called central exponent $\Omega(A)$, was constructed in [12], p. 114. The attainability of this estimate was proved in [47] with the use of the classical rotation method. This problem was solved in [31] and [30] for linear systems with perturbations decreasing at infinity at various rates and in [5] for linear systems with perturbations determined by integral conditions. Later in [43] and [44] perturbations infinitesimal in mean with a weight function have been investigated. The recent monograph [32] is almost completely devoted to this problem.

3 Bounded Perturbation

In this chapter we will consider the perturbation set

$$\mathfrak{M}_q = \{\Delta = (\Delta(n))_{n\in\mathbb{N}} : \|\Delta\|_\infty < q\}, \tag{7}$$

where $\|\Delta\|_\infty = \sup \|\Delta(n)\|$. In the next paragraph we present a definition of stability of Lyapunov exponents and a sufficient condition for the stability. Next, we will present analytic formulas for maximal upper and minimal lower movability boundaries of the higher exponent of (1) with perturbation in the class $\mathfrak{M}_q$ in two-dimensional stationary case. After that, we will present basic facts about generalized spectral radius and finally, we will show how this tool may be used to determine $\Lambda_u(\mathfrak{M}_q)$ and $\Lambda_u(\mathfrak{M}_q)$ in case of stationary system.

We have the following definition.

Definition 3. The Lyapunov exponents of system (1) are called stable if for any $\varepsilon > 0$ there exists $\delta > 0$ such that the inequality

$$\sup_{n\in N} \|\Delta(n)\| < \delta \tag{8}$$

implies the inequality

$$\left|\lambda_i'(A) - \lambda_i'(A+\Delta)\right| < \varepsilon,\ i = 1,...,s.$$

To formulate our main results for a Green's matrix of (1) denote by $x_i(m,n)$ the i-th column of it and by μ_i the characteristic exponent of the sequence $(\|x_i(m,n)\|)_{m\in N}$, $i = 1,...,s$. The next theorem [18] constitutes discrete time version of Malkin [45] sufficient condition for continuity of Lyapunov exponents.

Theorem 4. *Suppose that for certain Green's $\mathcal{G}(m,n)$ matrix of (1) and any $\gamma > 0$ there exists $d > 0$ such that*

$$\|x_i(m,n)\| \le d\exp[(\mu_i+\gamma)(m-n)] \quad \textit{for } m,n\in N,\ m\ge n,\ i=1,...,s \tag{9}$$

and

$$\|x_i(m,n)\| \le d\exp[(\mu_i - \gamma)(m-n)] \;\; \textit{for } m,n \in N,\ n \ge m,\ i = 1,...,s, \tag{10}$$

then the Lyapunov exponents of system (1) are stable.

Using Theorem 4 it can be shown that the Lyapunov exponents of time-invariant system are stable for invertible matrix system.

Theorem 5. *Lyapunov exponents of time-invariant system*

$$x(n+1) = Ax(n), \tag{11}$$

with invertible matrix A are stable.

Consider a time-invariant two-dimensional system

$$x(n+1) = Ax(n), n \ge 0, \tag{12}$$

where A is a two-by-two matrix in Jordan canonical form

$$A = \begin{bmatrix} a_1 & 0 \\ 0 & a_2 \end{bmatrix} \tag{13}$$

or

$$A = \begin{bmatrix} a_1 & 0 \\ 1 & a_2 \end{bmatrix}, \tag{14}$$

where a_1, a_2 are positive. Together with (12) we consider the following disturbed system

$$y(n+1) = (A + Q(n))\,y(n), \tag{15}$$

where $(Q(n))_{n\in\mathbb{N}}$,

$$Q(n) = \begin{bmatrix} q_{11}(n) & q_{12}(n) \\ q_{21}(n) & q_{22}(n) \end{bmatrix}$$

is a sequence of two-by-two matrices such that

$$|q_{ij}(n)| \le q \tag{16}$$

for all $i,j = 1,2$ and all $n = 0,1,\ldots$.

In the considered case the set of Lyapunov exponents of system (15) contains at most 2 elements, say $\lambda_1(A+Q) \le \lambda_2(A+Q)$. We will try to describe the influence of the perturbation $(Q(n))_{n\in\mathbb{N}}$, on the greatest Lyapunov exponent of (12). We will investigate the following counties $\lambda_2^{\min}(A,q) = \min\lambda_2(A+Q)$, $\lambda_2^{\max}(A,q) = \max\lambda_2(A+Q)$,where the maxima and minima are taken over all perturbation sequences $(Q(n))_{n\in\mathbb{N}}$ satisfying (16).

The following Theorem contains analytic expressions for $\lambda_2^{\max}(A,q)$.

Theorem 6. *We have*

$$\lambda_2^{\max}(A,q)=\begin{cases}\ln\left[\frac{1}{2}\left(a_1+a_2+2q+\sqrt{(a_2-a_1)^2+4q^2}\right)\right] \\ \text{if } A \text{ is given by (13)} \\ \ln\left[\frac{1}{2}\left(a_1+a_2+2q+\sqrt{(a_2-a_1)^2+4q(1+q)}\right)\right] \\ \text{if } A \text{ is given by (14)}\end{cases}.$$

The problem of evaluating $\lambda_2^{\min}(A,q)$ seems to be much harder. Some particular cases are provided in the next Theorem.

Theorem 7. *If A is given by (13) and $4q \leq |a_2-a_1|$, then we have*

$$\lambda_2^{\min}(A,q)= \ln\frac{a_1+a_2+\sqrt{(|a_2-a_1|-4q)\,|a_2-a_1|}}{2}.$$

4 Generalized Spectral Radius and Subradius

Denote by $\rho(A)$ the spectral radius of a matrix A. Consider a nonempty set Σ of s-by-s matrices. For $m\geq 1$, Σ^m is the set of all products of matrices in Σ of length m,

$$\Sigma^m=\{A_1A_2...A_m : A_i\in\Sigma, i=1,...,m\}.$$

Set

$$\overline{\alpha}_m=\sup_{A\in\Sigma^m}\|A\|,\ \underline{\alpha}_m=\inf_{A\in\Sigma^m}\|A\|,$$

$$\overline{\beta}_m=\sup_{A\in\Sigma^m}\rho(A),\ \underline{\beta}_m=\inf_{A\in\Sigma^m}\rho(A)$$

and define:

- the joint spectral subradius

$$\widehat{\rho}_*(\Sigma)=\inf_{m\geq 1}\underline{\alpha}_m^{1/m},$$

- the joint spectral radius

$$\widehat{\rho}(\Sigma)=\inf_{m\geq 1}\overline{\alpha}_m^{1/m},$$

-the generalized spectral subradius

$$\overline{\rho}_*(\Sigma)=\inf_{m\geq 1}\underline{\beta}_m^{1/m},$$

-the generalized spectral radius

$$\overline{\rho}(\Sigma)=\sup_{m\geq 1}\overline{\beta}_m^{1/m}.$$

The concepts of joint and generalized spectral radii were introduced in [53] and in [20] (see also [21]), respectively. Next in [8] and [23] two different proofs of the equality

$$\widehat{\rho}(\Sigma) = \overline{\rho}(\Sigma) \tag{17}$$

were given for the bounded set Σ. In [20] it was also shown that for bounded set Σ we have

$$\widehat{\rho}(\Sigma) = \lim_{m\to\infty} \overline{\alpha}_m^{1/m} = \limsup_{m\to\infty} \overline{\beta}_m^{1/m}. \tag{18}$$

Later for bounded set Σ we will denote the common value of $\widehat{\rho}(\Sigma)$ and $\overline{\rho}(\Sigma)$ by $\rho(\Sigma)$. The concepts of joint and the generalized spectral subradii were introduced in [26] to present conditions for Markov asymptotic stability of a discrete linear inclusion. In this paper it has been also shown that

$$\widehat{\rho}_*(\Sigma) = \overline{\rho}_*(\Sigma) \tag{19}$$

for finite Σ. In [9] these concepts have been related to the so called mortality problem. We say that the set of matrices Σ is mortal if the zero matrix can be expressed as the product of finitely many matrices from Σ. It appears that Σ is mortal if, and only if, $\widehat{\rho}_*(\Sigma) = 0$. Finally, in [13] inequality (19) was extended to the case of any nonempty set of matrices and it was shown that

$$\widehat{\rho}_*(\Sigma) = \lim_{m\to\infty} \underline{\alpha}_m^{1/m} = \liminf_{m\to\infty} \underline{\beta}_m^{1/m}. \tag{20}$$

Later for nonempty set Σ we will denote the common value of $\widehat{\rho}_*(\Sigma)$ and $\overline{\rho}_*(\Sigma)$ by $\rho_*(\Sigma)$. Because of the equalities (17) and (19) we can introduce the following definition.

Definition 4. For bounded set Σ we will denote the common value of $\widehat{\rho}(\Sigma)$ and $\overline{\rho}(\Sigma)$ by $\rho(\Sigma)$ and called it generalized spectral radius. For nonempty set Σ we will denote the common value of $\widehat{\rho}_*(\Sigma)$ and $\overline{\rho}_*(\Sigma)$ by $\rho_*(\Sigma)$ and called it generalized spectral subradius.

Denote by $\mathscr{D}(\Sigma)$ the set of all infinite sequences of elements of Σ. For fixed $d \in \mathscr{D}(\Sigma)$, $d = (A(1), A(2), \ldots)$ define $\Phi_d(m) = A(m-1)\ldots A(1)A(0)$ and

$$\overline{\rho}(d) = \limsup_{m\to\infty} \|\Phi_d(m)\|^{\frac{1}{m}}.$$

From the definitions of $\rho(\Sigma)$, $\rho_*(\Sigma)$ and $\overline{\rho}(d)$ the following inequality follows

$$\rho_*(\Sigma) \le \overline{\rho}(d) \le \rho(\Sigma)$$

for bounded set Σ. Much deeper relations between these three quantities is given by the next Theorem.

Theorem 8. *For any nonempty set* Σ *we have*

$$\rho_*(\Sigma) = \inf_{d \in \mathscr{D}(\Sigma)} \overline{\rho}(d). \quad (21)$$

For any nonempty and bounded set $\sum$ *we have*

$$\rho(\Sigma) = \sup_{d \in \mathscr{D}(\Sigma)} \overline{\rho}(d) \quad (22)$$

and if Σ *is in addition closed then the sup is max. Moreover, if the matrices in nonempty and bounded* Σ *are invertible, then for each* $\gamma \in (\rho_*(\Sigma), \rho(\Sigma))$ *there exists* $d \in \mathscr{D}(\Sigma)$ *such that* $\overline{\rho}(d) = \gamma$.

Equalities (21) and (22) have been proved in [13] and [23], respectively. The attainability of sup in (22) has been established in [20] for finite Σ and next in [54] this result has been extended to the case of compact set. The last statement of the Theorem is shown in [15].

Unfortunately, if the family Σ is not just a single matrix, the computation of $\rho(\Sigma)$ and $\rho_*(\Sigma)$ are not easy tasks at all. The problem of numerical computation of $\rho(\Sigma)$ and $\rho_*(\Sigma)$ is discussed in [9]-[11], [24], [25] and [42]; see also the references therein.

5 Central Exponents: Definitions and Basic Properties

Using the concept of central exponents of families of sequences, we now define the concept of upper and lower central exponents of system (12).

Definition 5. The upper (lower) sequence of (1) is upper (lower) sequence of the family

$$\left\{ \left(\ln \frac{\|x(n+1,x_0)\|}{\|x(n,x_0)\|} \right)_{n \in \mathbb{N}} : x_0 \in R^s, \|x_0\| = 1 \right\}. \quad (23)$$

The set of all upper (lower) sequences of (1) will be denoted by $\mathscr{U}(A)$ ($\mathscr{L}(A)$). Analogically, upper (lower) central exponent of (1) is defined as upper (lower) central exponent of (23) and will be denoted as $\Omega(A)$ ($\omega(A)$).

Notice that the definition is correct. One may obtain the following characterization of upper and lower sequences in terms of transition matrix.

Theorem 9. *Bounded sequences* $(r(n))_{n \in \mathbb{N}}$ *and* $(R(n))_{n \in \mathbb{N}}$ *are lower and upper sequences for (1), respectively, if and only if, for any* $\varepsilon > 0$ *there exist constants* $d_{r,\varepsilon}$ *and* $D_{R,\varepsilon}$ *such that for all* $m > k$ *we have*

$$d_{R,\varepsilon} \exp\left(\sum_{i=k}^{m-1} (r(i) - \varepsilon) \right) \leq \|\mathscr{A}(m,k)\| \leq D_{R,\varepsilon} \exp\left(\sum_{i=k}^{m-1} (R(i) + \varepsilon) \right). \quad (24)$$

The lower central exponent does not require special consideration since the problem can be reduced to the investigation of upper central exponent for the adjoint system. This is the content of the next theorem.

Theorem 10. *The lower central exponent ω of (1) is equal to the upper central exponent of the adjoint system (5), taken with the opposite sign.*

We have also the following formulas for upper and lower central exponent of (1) in terms of transition matrix.

Theorem 11. *There exist limits*

$$\lim_{N\to\infty}\frac{1}{N}\left(\overline{\lim_{n\to\infty}}\frac{1}{n}\sum_{i=0}^{n-1}\ln\|\mathscr{A}(N+i,i)\|\right)$$

$$\lim_{N\to\infty}\frac{1}{N}\left(\overline{\lim_{n\to\infty}}\frac{1}{n}\sum_{i=0}^{n-1}\ln\|\mathscr{A}((i+1)N,iN)\|\right),$$

and they are equal to

$$\Omega(A)=\inf_{N\in\mathbb{N}}\frac{1}{N}\left(\overline{\lim_{n\to\infty}}\frac{1}{n}\sum_{i=0}^{n-1}\ln\|\mathscr{A}(N+i,i)\|\right)$$

$$=\inf_{N\in\mathbb{N}}\frac{1}{N}\left(\overline{\lim_{n\to\infty}}\frac{1}{n}\sum_{i=0}^{n-1}\ln\|\mathscr{A}((i+1)N,iN)\|\right).$$

By Theorem 10 one can obtain analogical formulas for lower central exponent. Observe first that the orthogonality of matrices $U(n)$ in Theorem 3 leads to the following Theorem.

Theorem 12. *With the notation of Theorem 3 we have*

$$\mathscr{U}(A)=\mathscr{U}(C),\ \mathscr{L}(A)=\mathscr{L}(C).$$

In particular it implies that the central exponents of $(A(n))_{n\in\mathbb{N}}$ and $(C(n))_{n\in\mathbb{N}}$ are equal.

Consider now system (1) with matrices $A(n)$ being upper triangular with diagonal elements $a_{ii}(n)$. Denote $A_d(n)=diag\,[a_{ii}(n)]_{i=1,\dots,s}$ and by $a_{ij}(n)$ and $z_{ij}(n,k)$, the elements of $A(n)$ and $\mathscr{A}(n+1,k)$, respectively. By a straightforward calculation we have that for $n>k$

$$z_{ij}(n,k)=\begin{cases}\sum\limits_{p=k}^{n-1}\sum\limits_{l=i+1}^{j}a_{il}(n-p)z_{lj}(n-p-1,k)\prod\limits_{q=n-p+1}^{n}a_{ii}(q)\ \text{for } i<j\\ \prod\limits_{q=m}^{n}a_{ii}(q)\ \text{for } i=j\\ 0\ \text{for } i>j\end{cases}. \tag{25}$$

The next Theorem describes relation between upper, lower sequences of the original system and those with matrix coefficients A_d.

Theorem 13. *We have* $\mathscr{U}(A) = \mathscr{U}(A_d)$, $\mathscr{L}(A) = \mathscr{L}(A_d)$, $\Omega(A) = \Omega(A_d)$ *and* $\omega(A) = \omega(A_d)$.

If $(A(n))_{n\in\mathbb{N}}$ and $\left(A^{-1}(n)\right)_{n\in\mathbb{N}}$ are bounded by a and a', respectively, then from definition of central and Lyapunov exponent is clear that

$$-\ln a' \leq \omega(A) \leq \lambda_s(A) \leq \lambda_g(A) \leq \Omega(A) \leq \ln a.$$

It is also not difficult to construct examples where $\Omega(A) < \ln a$, $\lambda_s(A) < \lambda_g(A)$ and $-\ln a' < \omega(A)$. There are known examples for which $\lambda_g(A) < \Omega(A)$.

Next theorem shows that under the condition that $\Delta(n) \to 0$, the central exponents of (1) and (6) coincide. Observe that invertibility of $A(n)$ implies that $A(n) + \Delta(n)$ are invertible starting from certain n_0. We will assume that they are invertible for all natural n and then the central exponents of (6) are well defined.

Theorem 14. *If* $\lim_{n\to\infty} \|\Delta(n)\| = 0$ *then* $\mathscr{L}(A) = \mathscr{L}(A+\Delta)$, $\mathscr{U}(A) = \mathscr{U}(A+\Delta)$ *and in particular* $\Omega(A) = \Omega(A+\Delta)$ *and* $\omega(A) = \omega(A+\Delta)$.

6 Regular Systems and Regularity Coefficients

In order to measure the irregularity of the system (1) some numerical characteristics, which are called coefficients of regularity are introduced. In this section we will consider three of them.

1. Lyapunov's coefficient of regularity ([40]) is defined as:

$$\sigma_L = \min \sigma_V - \liminf_{n\to\infty} \frac{1}{n} \ln |\det \mathscr{A}(n,0)|,$$

where minimum is taken over the set of all bases. In fact it is enough to take the minimum over the set of normal bases.

2. Perron's coefficient of regularity ([49]). Consider the values

$$\lambda_1' \leq \lambda_2' \leq \ldots \leq \lambda_s' \tag{26}$$

and

$$\mu_s' \leq \mu_{s-1}' \leq \ldots \leq \mu_1' \tag{27}$$

of the Lyapunov exponents of (1) and (5), respectively, counted with their multiplicities. Then Perron's coefficient of regularity is defined as

$$\sigma_P = \max_{i=1,\ldots,s} \left(\lambda_i' + \mu_i'\right).$$

3. Grobman's coefficient of regularity ([12]). For a pair of dual bases $V = \{v_1, ..., v_s\}$ and $W = \{w_1, ..., w_s\}$ we define defect of dual bases

$$\gamma(V,W) = \max_{i=1,...,s} (\lambda(v_i) + \mu(w_i)) .$$

Then Grobman's coefficient of regularity is defined as:

$$\sigma_G = \min \gamma(V,W), \tag{28}$$

where the minimum is taken over all pairs of dual bases.

We will also say about regularity of the sequence $(A(n))_{n\in\mathbb{N}}$ instead of regularity coefficient of (1).

The introduced coefficients σ_P, σ_L and σ_G are related by the following inequalities

$$0 \le \sigma_P \le \sigma_G \le s\sigma_P \tag{29}$$

and

$$0 \le \sigma_G \le \sigma_L \le s\sigma_G. \tag{30}$$

(see, [7] Theorem 1.2.6 for the proof of (29) and [16] Lemma 1 for the proof of (30)). It appears that considering regularity coefficients we may restrict ourselves to the uppertriangular system according to the following Theorem from [7].

Theorem 15. *If sequence $(C(n))_{n\in\mathbb{N}}$ is constructed for sequence $(A(n))_{n\in\mathbb{N}}$ according to Theorem 3, then the regularity coefficients σ_P, σ_L and σ_G are the same for $(A(n))_{n\in\mathbb{N}}$ and $(C(n))_{n\in\mathbb{N}}$.*

The next Theorem contains the main result of this section.

Theorem 16. *If*

$$\lambda(\Delta) < -\sigma_G , \tag{31}$$

then the spectra of (1) and (6) coincide.

As it is shown in the next theorem, in case of diagonal matrices $A(n)$, spectra of (1) and (6) coincide for perturbations with characteristic exponent equal to $-\sigma_G$.

Theorem 17. *If the matrices $A(n)$ are diagonal and*

$$\lambda(\Delta) \le -\sigma_G < 0, \tag{32}$$

then the spectra of (1) and (6) coincide.

Together with (1) consider the following non-homogeneous system

$$x(n+1) = A(n)x(n) + f(n), \tag{33}$$

where the sequence $f = (f(n))_{n\in\mathbb{N}}$ belongs to the class F consisting of all sequences g of s-dimensional vectors such that

$$-\infty < \lambda(g) < \infty.$$

For an initial condition x_0 the solution of (33) is denoted by $x(n, x_0)$ so

$$x(n,x_0,f) = \mathscr{A}(n,0)x_0 + \sum_{i=0}^{n-1} \mathscr{A}(n,i+1)f(i). \tag{34}$$

From this formula it follows that the set $\left\{\lambda\left((x(n,x_0,f))_{n\in\mathbb{N}}\right) : x_0 \in R^s, x_0 \neq 0\right\}$ contains at most $s+1$ elements. Denote by $\chi(A,f)$ the minimal characteristic exponents of solution of (33) that is

$$\chi(A,f) = \min_{x_0\neq 0} \lambda\left((x(n,x_0,f))_{n\in\mathbb{N}}\right). \tag{35}$$

We introduce a quantity $\sigma(A)$ which will measure the difference between Lyapunov exponents of the non-homogeneous system (33) and of f. Define $\sigma(A)$ by the following formula

$$\sigma(A) = \sup_{f\in F}\left(\chi(A,f) - \lambda(f)\right). \tag{36}$$

Finally, we have the following result.

Theorem 18. *The following inequality holds*

$$\frac{\sigma_G}{s} \leq \sigma(A) \leq \sigma_G. \tag{37}$$

In particular, system (1) is regular if, and only if, $\sigma(A) = 0$.

References

1. Amato, F., Mattei, M., Pironti, A.: A note on quadratic stability of uncertain linear discrete-time systems. IEEE Trans. Automat. Control 43(2), 227–229 (1998)
2. Amato, F., Mattei, M., Pironti, A.: A robust stability problem for discrete-time systems subject to an uncertain parameter. Automatica 34(4), 521–523 (1998)
3. Amato, F., Ambrosino, R., Ariola, M., Merola, A.: A Procedure for Robust Stability Analysis of Discrete-Time Systems via Polyhedral Lyapunov Functions. In: Proceedings of the 2008 American Control Conference, Seattle, Washington, USA (2008)
4. Arnold, L., Pierre Eckmann, J., Crauel, H. (eds.): Lyapunov Exponents, Proceedings of a Conference, held in Oberwolfach, Germany, May 28-June 2, 1990. Lecture Notes in Mathematics, vol. 1486. Springer, Berlin (1991)
5. Barabanov, E.A., Vishnevskaya, O.G.: Exact bounds of Liapunov's exponents for linear differential perturbed systems with integrally restricted perturbation matrices on the semiaxis. Dokl. Akad. Nauk Belarusi 41(5), 29–34 (1997)
6. Barreira, L., Pesin, Y.: Lyapunov exponents and Smooth Ergodic Theory. Univ. Lecture Ser., vol. 23. Amer. Math Soc., Providence (2002)
7. Barreira, L., Valls, C.: Stability theory and Lyapunov regularity. Journal of Differential Equations 232(2), 675–701 (2007)
8. Berger, M.A., Wang, Y.: Bouned semigroups of matrices. Linear Algebra Appl. 166, 21–27 (1992)
9. Blondel, V.D., Tsitsiklis, J.N.: When is a pair of matrices mortal? Inform. Process. Lett. 63, 283–286 (1997)

10. Blondel, V.D., Tsitsiklis, J.N.: Complexity of stability and controllability of elementary hybrid systems. Automatica 35, 479–489 (1999)
11. Blondel, V.D., Tsitsiklis, J.N.: Boundedness of all products of a pair of matrices is undecidable. Systems and Control Letters 41(2), 135 (2000)
12. Bylov, B.F., Vinograd, R.E., Grobman, D.M., Nemytskii, V.V.: Theory of Lyapunov Exponents and Its Applications to Stability Theory, Moscow, Nauka (1966) (in Russian)
13. Czornik, A.: On the generalized spectral subradius. Linear Algebra and Its Applications 407(1-3), 242–248 (2005)
14. Czornik, A., Jurgas, P.: Some properties of the spectral radius of a set of matrices. International Journal of Applied Mathematics and Computer Science 16(2), 183–188 (2006)
15. Czornik, A., Jurgas, P.: Set of possible values of maximal Lyapunov exponents of discrete time-varying linear system. Automatica 44(2), 580–583 (2008)
16. Czornik, A., Nawrat, A.: On the regularity of discrete linear systems. Linear Algebra and Its Applications 432(11), 2745–2753 (2010)
17. Czornik, A., Nawrat, A.: On the Sigma Exponent of Discrete Linear Systems. IEEE Transactions on Automatic Control 55(6), 1511–1515 (2010)
18. Czornik, A., Nawrat, A., Niezabitowski, M.: On the stability of Lyapunov exponents of discrete linear systems. Submitted for publication in IEEE Transactions on Automatic Control (2012)
19. Dai, X.: Exponential stability of nonautonomous linear differential equations with linear perturbations by Liao methods. Journal of Differential Equations 225, 549–572 (2006)
20. Daubechies, I., Lagarias, J.C.: Sets of matrices all infinite products of which converge. Linear Algebra Appl. 161, 227–263 (1992)
21. Daubechies, I., Lagarias, J.C.: Corrigendum/addendum to: Sets of matrices all infinite products of which converge. Linear Algebra Appl. 327, 69–83 (2001)
22. Demidovich, V.B.: Stability criterion for difference equations. Diff. Uravneniya 5(7), 1247–1255 (1969) (in Russian)
23. Elsner, L.: The generalized spectral radius theorem: An analytic-Geometric proof. Linear Algebra Appl. 220, 151–159 (1995)
24. Gripenberg, G.: Computing the join spectral radius. Linear Algebra Appl. 234, 43–60 (1996)
25. Guglielmi, N., Wirth, F., Zennaro, M.: Complex Polytope Extremality Results for Families of Matrices. Siam J. Matrix Anal. Appl. 27(3), 721–743 (2005)
26. Gurvits, L.: Stability of discrete linear inclusion. Linear Algebra Appl. 231, 47–85 (1995)
27. Hinrichsen, D., Prichard, A.J.: Stability radii of Linear Systems. Systems and Control Letters 7, 1–10 (1986)
28. Hinrichsen, D., Prichard, A.J.: Mathematical systems theory. I. Applied Mathematics, vol. 48. Springer, Berlin (2005)
29. Izobov, N.A.: Exponential Stability by the Linear Approximation. Differential Equations 37(8) (2001)
30. Izobov, N.A.: The Higher Exponent of a Linear System with Exponential Perturbations. Differential Equations 5(7), 1186–1192 (1969)
31. Izobov, N.A.: The Higher Exponent of a System with Perturbations of Order Higher Than One. Vestn. Bel. Gos. Unta. Ser. I (3), 6–9 (1969) (in Russian)
32. Izobov, N.A.: Vvedenie v teoriyu pokazatelei Lyapunova (Introduction to the Theory of Lyapunov Exponents). Belarus. Gos. Univ., Minsk (2006) (in Russian)
33. Jury, E.I.: Robustness of a discrete system. Automation and Remote Control 51, 571–592 (1990)
34. Kolla, S.R., Yedavalli, R.K., Farison, J.B.: Robust stability bounds on time-varying perturbations for state-space models of linear discrete-time systems. International Journal of Control 50, 151–159 (1989)
35. Kryzhevich, S.G.: Estimates of Exponents of Linear Nonhomogeneous Systems via the Grobman Improperness Coefficient. Differential Equations 38(8), 1208–1210 (2002)

36. Kuznetsov, N.V., Leonov, G.A.: Counterexample of Perron in the Discrete Case. Izv. RAEN, Diff. Uravn. 5, 71 (2001)
37. Kuznetsov, N.V., Leonov, G.A.: Stability Criteria by the First Approximation for Discrete Nonlinear Systems (part I), Vestnik St. Petersburg University. Mathematics 39(2), 55–63 (2005)
38. Kuznetsov, N.V., Leonov, G.A.: Stability Criteria by the First Approximation for Discrete Nonlinear Systems (part II), Vestnik St. Petersburg University. Mathematics 39(3), 30–42 (2005)
39. Lima, R., Rahibe, M.: Exact Lyapunov exponent for infinite products of random matrices. Journal of Physics A: Mathematical and General 27(10), 3427–3437 (1994)
40. Lyapunov, A.M.: General problem of stability of motion, Collected works. Izdat. Akad. Nauk SSSR 2 (1956) (in Russian)
41. Leonov, G.A.: Lyapunov exponents and problems of linearization. From stability to chaos. St. Petersburg University Press (1997)
42. Maesumi, M.: Calculating joint spectral radius of matrices and Holder exponent of wavelets. In: Chui, C.K., Schumaker, L.L. (eds.) Approximation Theory IX. World Scientific, Singapore (1988)
43. Makarov, E.K., Marchenko, I.V., Semerikova, N.V.: On an Upper Bound for the Higher Exponent of a Linear Differential System with Integrable Perturbations on the Half-Line. Differents. Uravn. 41(2), 215–224 (2005)
44. Makarov, E.K., Marchenko, I.V.: On an Algorithm for Constructing an Attainable Upper Boundary for the Higher Exponent of Perturbed Systems. Differential Equations 41(12), 1621–1634 (2005)
45. Malkin, I.G.: Theory of stability of motion, Nauka, Moscow (1966) (in Russian)
46. Mane, R.: Lyapunov exponents and stable manifolds for compact transformations. In: Geometric Dynamics. Lecture Notes in Math., vol. 1007, pp. 522–577. Springer (1983)
47. Millionshchikov, V.M.: A proof of attainability of central exponents of linear systems. Sibirskii Matematicheskii Zhurnal 10, 99–104 (1969)
48. Molchanov, A.P., Liu, D.: Robust Absolute Stability of Time-Varying Nonlinear Discrete Systems. IEEE Transactions on Automatic Control 49, 1129–1137 (2002)
49. Perron, O.: Über Stabilität und asymptotisches Verhalten des Integrale von Differentialgleichungssystemen. Mathematische Zeitschrift 31, 748–766 (1929)
50. Perron, O.: Die Stabilitätsfrage bei Differentialgleichungen. Mathematische Zeitschrift 32(5), 702–728 (1930)
51. Gyori, I., Pituk, M.: The Converse of the Theorem on Stability by the First Approximation for Difference Equations. Nonlinear Analysis 47, 4635–4640 (2001)
52. Phat, V.N.: On the stability of time-varying systems. Optimization 45, 237–254 (1999)
53. Rota, G.C., Strang, G.: A note on the joint spectral radius. Inag. Math. 22, 379–381 (1960)
54. Shen, J.: Compactification of a set of matrices with convergent infinite products. Linear Algebra Appl. 311, 177–186 (2000)
55. Su, J.H., Fong, I.K.: New robust stability bounds of linear discrete-time systems with time-varying uncertainties. International Journal of Control 58, 1461–1467 (1993)
56. Wang, K., Michel, A.N., Liu, D.: Necessary and sufficient conditions for the Hurwitz and Schur stability of interval matrices. IEEE Transactions on Automatic Control 39, 1251–1255 (1994)
57. Wirth, F.: On the calculation of real time-varying stability radii. International Journal of Robust and Nonlinear Control 8, 1043–1058 (1998)

Single DOF Powered Exoskeleton Control System, Algorithms and Signal Processing

Michał A. Mikulski

Abstract. The dynamic growth of robotics in the recent years has led to the creation of many machines directly interacting with the human body. An example of such devices are exoskeletons, a group of wearable robots. The main reason behind exoskeleton development is the augmentation of the physical abilities of a human being, specifically strength and endurance for the current state of the art. This article describes the most general categorization of exoskeletons, based on their control systems with the examples of commercially existing prototypes. The purpose of this article is to analyze the validity of using electromyogram control on a single degree-of-freedom powered exoskeleton using linear electric actuators. The primary effort of the research done was to create a control algorithm for the exoskeleton constructed. Due to the users individual properties, as well as other dynamically changing parameters, such as the skin conductance level or the electrode placement, a specific EMG circuit and calibration method was designed. With the signal unified, the dual channel differential threshold algorithm was created, based on the muscle tremor statistical analysis. The article describes these aspects of the exoskeleton research.

Keywords: electromyogram, biorobotics, powered exoskeleton, assistive technologies.

1 Introduction

By definition of J. L. Pons from [1] wearable robots can be described as person-oriented robots: worn by human operators to supplement or replace the function of

Michał A. Mikulski
Silesian University of Technology, Institute of Automatic Control,
Akademicka 16, 44-101 Gliwice, Poland
e-mail: michal.mikulski@polsl.pl

A. Nawrat et al. (Eds.): Advanced Technologies for Intelligent Systems, SCI 440, pp. 45–55.
springerlink.com

a limb. The main purpose of the technology behind wearable robots is to extend, complement, substitute or enhance a limb function beyond its natural abilities. Pons differentiates wearable robots into the following categories:

- Empowering robotic exoskeletons,
- Orthotic robots,
- Prosthetic robots.

Fig. 1 Single Degree-of-Freedom Upper Limb Powered Exoskeleton

For the time of the publication exoskeletons worldwide in production or advanced stages of development exist in either medical or military applications. For rehabilitation purposes exoskeletons such as Cyberdyne's HAL-5 [2, 3] or Berkley Bionics' eLegs [4] use biosignals to detect the user's movement intent, and drive the actuators accordingly. In defense and warfare applications Lockheed Matrin's HULC [5] and Raytheon's XOS2 [6] use force control to allow a wider range of movements for the user.

Due to the large interest in the assistive technologies worldwide a single degree-of-freedom powered exoskeleton controlled by electromyogram for the human upper limb was created at the Silesian University of Technology, presented in Fig. 1. This paper is a summary of the research done in electromyogram processing and control algorithms for powered exoskeletons.

2 Wearable Robots' Control Systems

With the different purposes of wearable robots and exoskeletons in particular comes the variance of control systems. The main reason is to allow the most intuitive control of the exoskeleton with the constraints given by the human operator. The main constraint on the control system is the state of the users motor function in the assisted limb.

2.1 Feed-Forward Systems

Exoskeletons used in rehabilitation and physiotherapy consider the fact that the user's limb motor functions can be impaired. That means that any sensory input collected from the user directly (e.g. electromyogram, force sensor, etc.) can have a substantial error due to user's disability. Such data cannot be used directly in a control algorithm, but can be used as a fuzzy data set in a feed forward loop. Gesture recognition is widely used in rehabilitation exoskeletons such as Cyberdyne's HAL-5 [2] and Berkley Bionics' eLegs [4]. This correlates to the fact, that those exoskeletons have a limited set of movements available. For example at the time of the publication of this article, eLegs exoskeleton allowed the user to do the following [4]:

- Walk in a straight line,
- Stand from a sitting position,
- Stand for an extended period of time,
- Sit down from a standing position.

Walking with an exoskeleton to disabled people is a significant improvement and currently a state-of-the-art technology. The full body rehabilitation exoskeletons require feed-forward systems to function properly. However the research on the single degree-of-freedom exoskeleton, presented in this paper, shows that for single limb physiotherapy, using an EMG signal based feedback control system, can be used with satisfactory results.

2.2 *Feedback Systems*

Exoskeletons used mainly in military applications such as Lockheed Martin's HULC [5], Raytheon's XOS2 [6] or Berkley Bionics' BLEEX [7] use feedback systems such as multi-axis foot force/torque sensors [8]. For any feedback system in full body robots the user must have fully functional limbs. Any disability, may influent the operation of a closed loop system. This directly corresponds to the available user movement type count. With the movements not limited by the gesture list of an open loop system, closed loop exoskeletons tend to work well in different environments. Single degree-of-freedom exoskeletons for rehabilitation can use closed loop systems, with biosignals such as the electromyogram, for training muscle groups with assistance from the wearable robot. Exoskeleton presented in this paper uses a feedback control system.

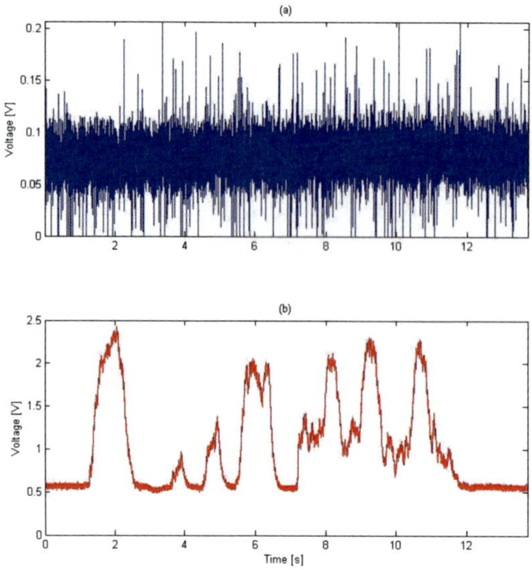

Fig. 2 EMG signal (a) after differential amplification stage 1 (b) after complete analog signal processing

3 EMG Processing

Electromyogram is an electric signal generated by the muscles due to flexion of muscle fibers. Because of its properties: low signal-to-noise ratio, low voltage (in comparison to current electronic devices) and high frequency, the EMG signal has to be filtered and amplified, before it can be used in a control system. Fig. 2 presents the EMG signal collected by the MCU from biceps brachii muscle during experimental runs.

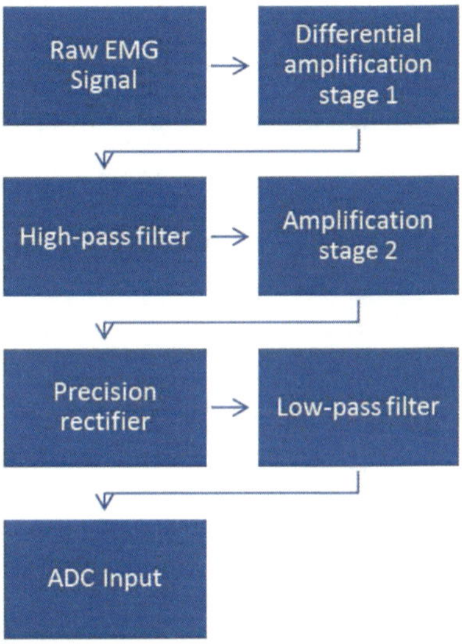

Fig. 3 Single channel EMG filter/amplifier diagram

The acquisition has been done using a 12bit ADC with voltage resolution of $805\mu V$. The signal after the first differential amplification, presented in Fig. 2 subfigure (a), cannot be used with the standard class, integrated ADC converters. To eliminate the need of using specialized SoC or DSP with advanced ADCs, an analog filter/amplifier circuit has been constructed, presented in Fig. 3. The goal of the circuit was to convert the high frequency EMG signal to a low frequency one, with a voltage level corresponding to a current muscle flexion.

The raw EMG signal is processed to eliminate the following conditions, that have a negative impact on the signal: voltage offset due to electrode placement, EMG cable state, noise from various sources such as skin conductance level, user's current physiological state or external electromagnetic field. The detailed analysis of electromyography can be found in [9, 10]. The final signal after the analog signal processing can be fed to a standard analog-to-digital converter located in the MCU. In this case, a Microchip dsPIC33FJ64MC504 Digital Signal Controller was used. Due to the low frequency of the final signal, the acquisition takes less CPU usage and doesn't use any additional digital signal processing.

Due to the individual features of a human being, such as the muscle tonus value, or the maximal muscle flexion, but also, due to some parameters already listed, responsible for voltage offset: electrode placement, physiological state, skin conductance, EMG channel calibration must be ran. Calibration procedure has been defined as a constant: 10 second time, during which absolute (minimal and maximal) EMG values are set. Those values are than used for unification of the EMG signal from absolute (displayed in voltage), to relative (0 being the lowest calibrated value, 1 being the highest). EMG calibration procedure is presented in Fig. 4.

Calibration procedure allows the user to choose a desirable contraction level, required to properly use the exoskeleton. The activation levels of the exoskeleton are set by flexing and relaxing the muscle groups during the calibration time. The values $\vartheta_{i_{calmin}}$ and $\vartheta_{i_{calmax}}$ are calculated by the equation stated below:

$$\bigwedge_{i=1}^{n} \vartheta_i > \vartheta_{i_{calmax}} \Rightarrow \vartheta_{i_{calmax}} := \vartheta_i \tag{1}$$

$$\bigwedge_{i=1}^{n} \vartheta_i < \vartheta_{i_{calmin}} \Rightarrow \vartheta_{i_{calmin}} := \vartheta_i \tag{2}$$

These values are responsible for residual muscle tension and maximal muscle flexion from the calibration time. Since they are the extreme values, they are hard to recreate during normal operations. Hence, for calculating relative contraction values, $\vartheta_{i_{calmin}}$ was magnified times 1.2 , and $\vartheta_{i_{calmax}}$ was reduced to times 0.8.

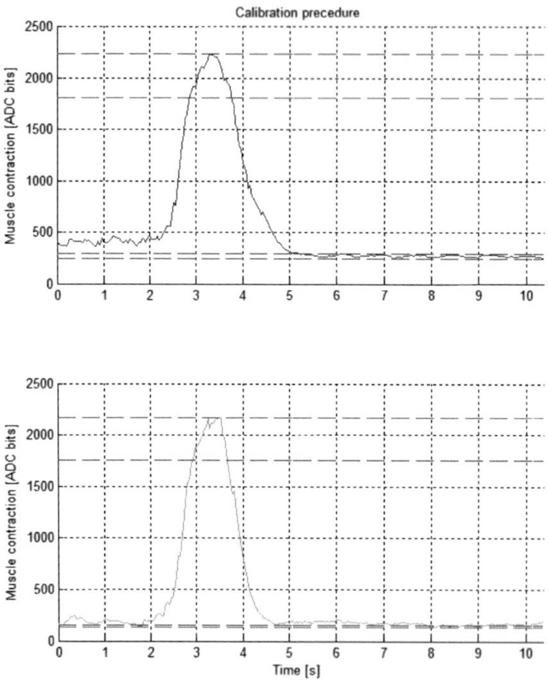

Fig. 4 Calibration procedure of two EMG channels: two EMG based signals (blue, green); boarder signal values (purple); boarder signal reduced values (red)

Furthermore, for the control algorithms, the current contraction values exceeding the calibration $\vartheta_{i_{calmin}}$ and $\vartheta_{i_{calmax}}$ are rounded up to those border values. Having the current relative muscle contraction level $\vartheta_i \in (0,1)$ simplifies the control algorithms later on, and prevent accidental over rotation.

4 Muscle Tremor Statistical Analysis

Exoskeletons can be used for muscle tremor suppression, described in detail in [11]. Exoskeleton presented in this paper can be used for tremor suppression with results depending on the algorithm used for control. While retaining a constant rotation angle of any limb, the muscles responsible for torque generation can twitch. A single channel proportional algorithm described in [12], without any additional anti-tremor suppression would fail in normal operations. Hence, statistical analysis of a muscle tremor is needed.

To determine muscle tremor statistical parameters, a user was asked to perform specific muscle flexion exercises. During the procedure, five regions where selected, in which the user tried to retain a steady muscle flexion value. The procedure is shown in Fig. 5.

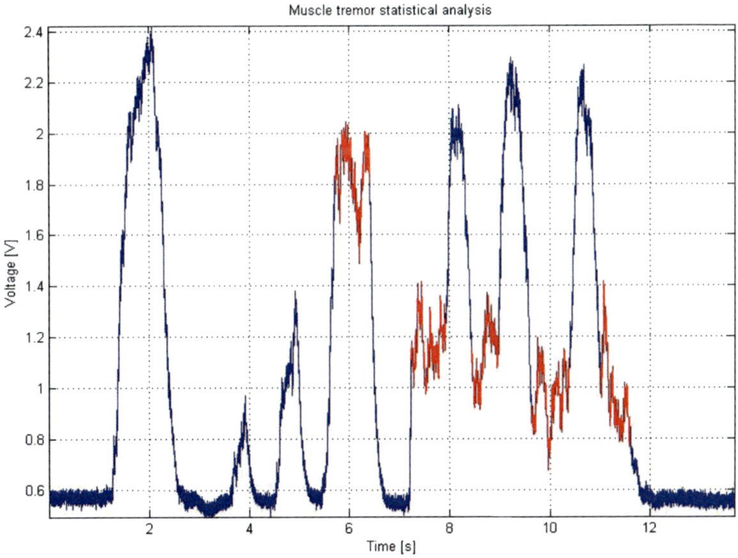

Fig. 5 Muscle tremor statistical analysis. Blue curve - absolute muscle contraction level. Red curves - periods where the user tried to retain a constant flexion value.

From the data collected for each period a standard deviation was calculated. The maximal standard deviation from all five periods is equal to $max\sigma_i = 0.142V$. Because the calibration method returned the minimal and maximal flexion/relaxation values $\vartheta_{i_{calmin}} = 0.4940V$ and $\vartheta_{i_{calmax}} = 2.052V$, that means that a typical muscle tremor ($max\sigma_i$) is no more than about 10% of the flexion value. We can assume that the muscle tremor distorts the EMG signal in a fashion of normal distribution, in which the mean value is corresponding to the current muscle flexion. Because all people are different, a confidence interval of 68% (equal to one $max\sigma_i$) may not be enough to estimate if the changing value is a result of movement or tremor. A minimal 30% deviation was set, to distinguish muscle tremor from movement.

5 EMG Control Algorithms

The constructed EMG signal processing allowed for the development of feedback control algorithms for a single degree-of-freedom powered exoskeleton. The mechanical structure of the exoskeleton was simplified and is presented in Fig. 6. A single rotary joint is located at the user's elbow, with the available rotation of $\Theta \in (\Theta_{min}, \Theta_{max}) = (15°, 90°)$.

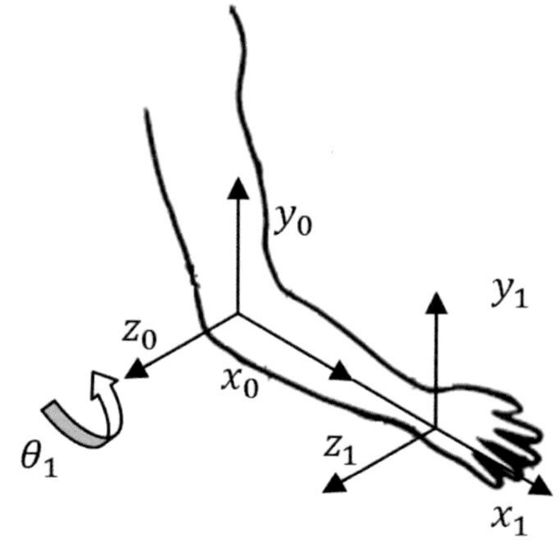

Fig. 6 Hartenberg-Denavit coordinates for Single-DOF exoskeleton arm

For the exoskeleton described in this article, three algorithms were developed:

- Single channel proportional algorithm,
- Single channel threshold algorithm,
- Dual channel differential threshold algorithm.

All of those algorithms are described in detail in [12]. The reason behind EMG processing and muscle tremor statistical analysis will be shown on Dual channel differential threshold algorithm. To achieve an intuitive control over the elbow join using an exoskeleton, the algorithm implements tension levels from biceps brachii (ϑ_1) and triceps brachii (ϑ_2) to calculate the differential muscle contraction e stated in eq. 3:

$$e = \vartheta_1 - \vartheta_2 \tag{3}$$

Since the biceps brachii and triceps brachii are an antagonist muscle pair, the signal e corresponds to the movement of the elbow. The exoskeleton in this research uses an electric linear actuator, meaning that the user can relax the muscle groups and the current rotation angle will be preserved only by the ball screw inside the actuator. This leads to the dual channel threshold algorithm stated with the equation (4), and presented in Fig. 7:

$$\Theta = \begin{cases} \Theta_{i-1} + d\Theta & \text{when } e > \sigma_f \\ \Theta_{i-1} & \text{when } e \in \langle \sigma_r, \sigma_f \rangle \\ \Theta_{i-1} - d\Theta & \text{when } e < \sigma_r \end{cases}, \tag{4}$$

In the equation σ_f and σ_r are flexion and relaxation thresholds calculated from the statistical tremor analysis. Having stated that a muscle group can have a standard

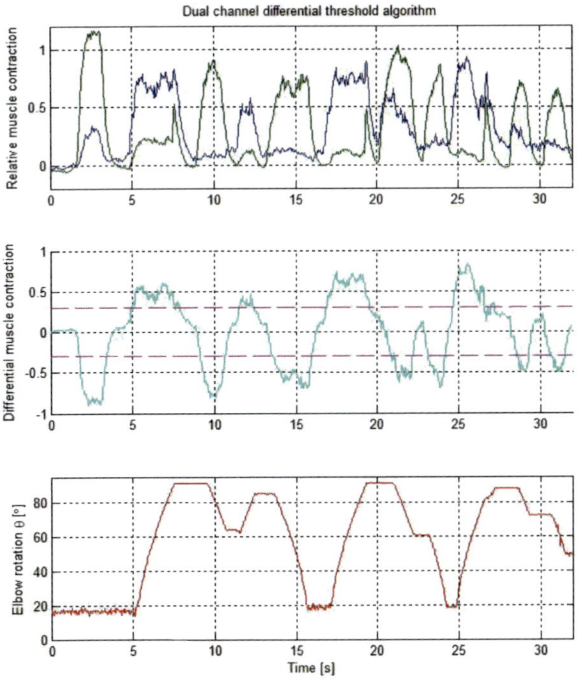

Fig. 7 Dual channel differential threshold algorithm

deviation up to 30%, $\sigma_f = 0.3$ and $\sigma_r = -0.3$. $d\Theta$ is the rotation increment between two work cycles.

Thresholds declared above allow a muscle tremor suppression, when the user's muscle contraction values have a similar relative value (no different then ± 0.3). When differential muscle contraction level $e > \sigma_f$, the exoskeleton forearm flexes the user's arm. When $e < \sigma_r$, the exoskeleton extends the forearm due to large influence of the triceps brachii. When the user contraction level is in between the thresholds $e \in \langle \sigma_r, \sigma_f \rangle$, joint rotation remains the same.

The dual channel differential threshold algorithm allows for intuitive control over a rotary joint, because of its biomechanical nature. Human limbs remain in a static equilibrium, when all forces and all torques are equal to zero. In non-assisted movement, the human body struggles with the force of gravity, to remain in position. Using a powered exoskeleton with electric linear actuation, the force of gravity is automatically canceled, by the actuator's holding torque. This allows for static equilibrium to be achieved, with differential muscle contraction level near zero. Also, the actuator's cancelation of gravity's pull allows for absolute muscle relaxation during exoskeletal control.

6 Conclusions

The electromyogram filter/amplifiers together with the feedback loop described in this article allowed for the creation of a single degree-of-freedom powered exoskeleton. The main purpose of this device is to assist users in rehabilitation and physiotherapy of the upper limb using biofeedback signals for both control and diagnosis. Electronic circuits and control algorithms for this exoskeleton structure can be transferred to a larger, more sophisticated construction of a lower limb or full body exoskeletons. During the experiments and testing of the exoskeleton's arm, a conclusion was reached, that the focus of biosignals in wearable robots must be on rehabilitation, rather than warfare. Due to unreliability of EMG in different situations and the need for calibration, when external conditions change (e.g. with the muscle fatigue, or change of skin conductance level) wearable robots using electromyogram signals may be unsuitable for quick movement and changing environment. Powered exoskeletons are currently an emerging topic in robotics. Many problems are yet to be resolved, starting with control methods for distinct applications, up to power consumption for long term usage.

References

1. Brunetti, F., Frizera, A., Pons, J.L., Bueno, L.: Human-robot cognitive interaction. In: Wearable Robots: Biomechatronic Exoskeletons, ch. 4, pp. 87–125. John Wiley & Sons Ltd, West Sussex (2008)

2. Lee, S., Kanabe, S., Sankai, Y., Kawamoto, H.: Power Assist Method for HAL-3 using EMG-based Feedback Controller. In: Proceedings of 2003 IEEE International Conference on Systems, Man and Cybernetics, Washington, D.C., USA, vol. 2, pp. 1648–1653 (2003)
3. Kasaoka, K., Sankai, Y.: Predictive control estimating operator's intention for stepping-up motion by exo-skeleton type power assist system HAL. In: Proceedings of International Conference on Intelligent Robots and Systems, Maui, HI, USA, vol. 3, pp. 1578–1583 (2001)
4. Berkley Bionics: About eLegs. Berkley Bionics Website, `http://berkeleybionics.com/exoskeletons-rehab-mobility/about-elegs/` (Cited June 30, 2011)
5. Lockheed Martin: HULC. Lockheed Martin's Website, `http://www.lockheedmartin.com/products/hulc/` (Cited June 30, 2011)
6. Lawrence, C., Marrapodi, E.: Future soldiers may be waring 'Iron Man' suits. CNN (2010)
7. Kazerooni, H., Zoss, A.B.: Biomechanical Design of the Berkley Lower Extremity Exoskeleton (BLEEX). IEEE/ASME Tranasactions on Mechatronics 11(2) (2006)
8. Lind, R.F., Love, L.J., Rowe, J.C., Pin, F.G.: Multi-axis foot reaction force/torque sensor for biomedical applications. In: Proceedings of the International Conference on Intelligent Robots and Systems, St. Louis, MO, USA (2009)
9. De Luca, C.J.: A Practicum on the Use of Surface EMG Signals in Movement Sciences (v1.5). DelSys Inc. (2007)
10. De Luca, C.J.: The Use of Surface Electromyography in Biomechanics. Journal of Applied Biomechanics 13, 135–163 (1997)
11. Rocon, E., Pons, J.L.: Exoskeletons in Rehabilitation Robotics. Springer, Berlin (2011)
12. Mikulski, M.: Electromyogram Control Algorithms for the Upper Limb Single-DOF Powered Exoskeleton. In: Proceedings of the 4th IEEE International Conference on Human System Interaction, Yokohama, Japan, pp. 117–122 (2011)

Part II
Design and Construction of Trajectory Planer for UAVs

In this chapter design and construction of trajectory planners are described. Control of UAVs can be multilevel. For instance, autonomous stabilization is one level of automation. Here, we discuss the problems with longer time of execution and at higher level of abstraction e.g. problem of planning of collision-free path for UAV. Usually, at the top of a typical, known from mobile robotics hierarchical, hybrid architecture there is a module called planer. It is used by a human operator that assigns various missions to the UAV by specifying the places (points) the vehicles should visit. The role of the planner is to generate the set of way-points that determine the collision-free path. The planner module is utilizing the information about the topology of the terrain. The contents of the chapter present a probabilistic approach to the problem of collision-free path planning in 3D space. It is presented how to acquire a feasible path in 3D space, simultaneously reducing the number of computations. The presented methods, thanks to the simplification, are both effective and require significantly reduced number of operations in comparison to full search.

Another important topic is cooperation of multiple robots. In the chapter it is presented how to design a solution of a specified problem in a multirobot environment. Presented solutions can be assumed as a popular benchmark in Multi-Agent Systems, which allows to verify different control, coordination and task distribution strategies. In order to make the examples more challenging an uncertainty was introduced and solved by a game theoretical approach. When knowledge about the environment is uncertain it is unwise to apply off-line planning. Here, an iteractive step-by-step execution of the example tasks is presented. Such a solution cannot guarantee optimal execution of the task. However, it was confirmed by a number of examples, that the presented approach is appropriate and the obtained solutions are acceptable. Another studied approach is based on probabilistic robot motion planning method called B-PPT built upon the PRM (Probabilistic Roadmap) method. The B-PPT algorithm is suitable for the problem of planning collision-free flight path for the UAV objects. The method is based on very simplified mathematical tools which use only necessary information. Therefore, it is fast and accurate but not repeatable, which reflects the human way of trajectory planning.

To conclude, the chapter presents various methods and algorithms of computation collision-free trajectory of flight for UAVs. The presented solutions are solved by classic, probabilistic, optimization and artificial intelligence approaches.

Probabilistic Approach to Planning Collision Free Path of UAV

Krzysztof Skrzypczyk, Adam Gałuszka, Marcin Pacholczyk, and Krzysztof Daniec

Abstract. In this paper an approach to planning collision free path based on probabilistic search is presented. The aim of this study is to design an algorithm for off-line planning a set of way-points which make the collision free route of an UAV. The planning process is based on the map that contains information about altitude of the terrain over which the UAV is intended to perform its mission. The probabilistic method of making the representative model of the terrain was applied in order to reduce a complexity of planning the collision free path. The functioning and efficiency of the approach proposed was illustrated with some exemplary simulations.

Keywords: probabilistic planning, UAV, surface models.

1 Introduction

The design and construction of Unmanned Aerial Vehicles (UAV) have become one of the most important and strategic branches of the contemporary Robotics. It is caused by a plenty of possible applications of this kind of automatons. Efficient and robust UAV are desired in many branches of industry. For instance in the military sector the UAV is a perfect tool for performing patrolling and reconnoitring missions. On the other hand in the civil sector of industry, unmanned vehicles are very useful for inspecting of large constructions as well as for monitoring, for instance, the traffic intensity in large cities. Of course there are many more potential and desired applications. Those aforementioned ones, are only examples which aim is to highlight an importance of this area of robotic research. Regarding the

Krzysztof Skrzypczyk · Adam Gałuszka · Marcin Pacholczyk · Krzysztof Daniec
Silesian University of Technology, Institute of Automatic Control,
Akademicka 16, 44-101 Gliwice, Poland
e-mail: {Krzysztof.Skrzypczyk,Adam.Galuszka}@polsl.pl,
{Marcin.Pacholczyk,Krzysztof.Daniec}@polsl.pl

A. Nawrat et al. (Eds.): Advanced Technologies for Intelligent Systems, SCI 440, pp. 59–68.
springerlink.com

construction of the UAV there are two main conceptions of the design. First one (relatively the simplest in realisation) is a remotely controlled vehicle. In this approach flying vehicle is controlled remotely by human operator. Using vision feedback (a camera mounted on board of the vehicle) the operator is able to move the flying object over the place or the region of interest and inspect it or make a photographic documentation. The second conception is the autonomous aerial vehicle. Such device is the most desired one, but simultaneously also the most difficult in designing and constructing. In this approach, the vehicle is intended to perform autonomously the mission assigned by the operator. It is easy to imagine how hard the problem of constructing such system is. During the process of the design, there are many serious and difficult problems to solve. It is enough to mention the collision free path planning, the vehicle stabilising in changing whether conditions, determining the aerial position and orientation and many others. Due to the a large number of possible applications and benefits related to them, intensive researches have been done in this area. [1, 13, 12, 15, 16, 18, 20]. This paper addresses the first of pointed out issues - the problem of determining the collision free path. Intended application of the developed UAV control system is performing autonomously the patrolling mission over an unstructured terrain. The goal of the planner is determining the set of way points that make the route the vehicle is intended to follow. There are many methods and algorithms of planning collision free paths, derived in a straight-line from mobile robotics [3, 4, 5, 6, 7, 8, 9, 13, 14, 17, 23]. In this work the probabilistic approach [2, 10, 11, 21] to solving the planning problem was selected and applied. The main reason of this choice was the demand of meeting main assumption which was the simplicity of the planning method as well as the minimalism of the representation required for the planning. Since the planning problem in the 3D space is relatively hard to solve in terms of complexity and costs of computations, the two-level approach to planning the three dimensional path was proposed. Together with the probabilistic method of building the model of the world representation such approach allows generating the collision free path with a small computational cost. The planning process is based on the topological map that contains information about the height of the terrain over which the vehicle is intended to perform its patrol mission. The work of the method is illustrated with an exemplary scenario that was modelled and simulated using MATLAB environment.

2 Overview of the System

In this section a brief overview of the navigation and control system is presented, in order to clearly address the framework in which presented problem is solved. Figure 1 shows general diagram of the navigation and control system of the UAV. It is created on the base of a typical, known from mobile robotics hierarchical, hybrid

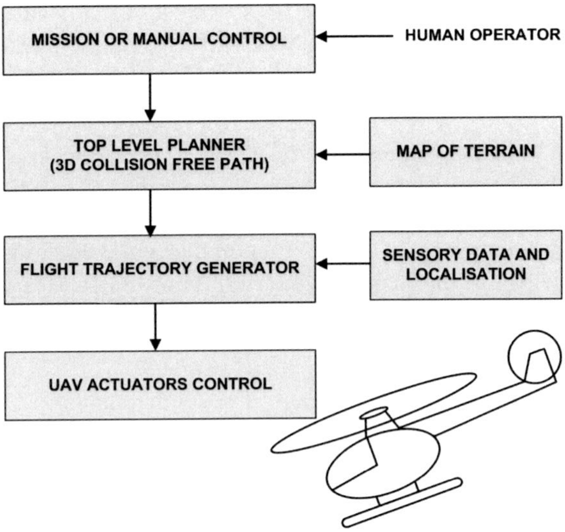

Fig. 1 The conceptual diagram of the UAV control system with the planner module being the point of the work presented

architecture. At the top level of the system, there is a module called *planner* which is the point of this study. The human operator assigns mission by specifying the places (points) the vehicle is intended to fly over. The role of the planner is to generate the set of way-points that determine the collision free path. The planner module is also provided with data about topology of the terrain. Using this data the model which is used further in the planning process is created. The next module is responsible for generating the collision free trajectory which enables the flight between the way-points.

The navigation module has to be also provided with information about the position and orientation in space. Determining the coordinates, and orientation is performed by the localisation subsystem. The plan created by the planner does not take into account the dynamics of the space inside of which the vehicle operates. Therefore the navigation system has to be also provided with data about existence of objects that can appear in the surrounding of the vehicle. This information is collected by the sensory system. The bottom level of this architecture is the motion control system which is to execute the trajectory generated by the navigation system. As it was mentioned before the core of this article is just the design and simulation evaluation of the algorithm dedicated to work with the planner module. The process of the design is presented and described in the rest part of this paper.

3 Problem Formulation

3.1 Assumptions

The planning system presented here is designed for the UAV which purpose is patrolling the frontiers areas. However it is created to satisfy a number of conditions and assumptions. They make some limitations for possible application and the framework in which the system work properly. So it is important to clearly point them out and discus. The following are the main assumptions that were followed during the design process:

- The patrol mission is performed on the given, priory stated height H_o.
- There is a map that provides information about height of the terrain points.
- The manoeuvre of avoiding obstacles is preferred to the flying over them

The assumptions pointed out, are the main limitations of the planning method presented. However they are realistic and justified and they do not limit the system to only simulation studies, and make the method applicable in real systems.

3.2 Problem Statement

Let us first define the mission to be performed by the UAV. The mission in fact is the collection of locations defined by human operator the vehicle is intended to fly over as well as the altitude H_0 of the flight. Let us denote the mission as the set:

$$M = \{m_1, m_2, ..., m_P\}, \quad m_i = (x_i, y_i),\ i = 1, 2, ..., P \tag{1}$$

The goal of the planner module is to calculate in the 3D space a set of points that lay between the locations defined by operator. Therefore for each pair of points:

$$(m_i, m_{i+1}),\ i = 1, 2, ..., P-1 \tag{2}$$

the module is intended to find a set of way points that determine the collision free path defined in 3D space:

$$S_i = \{s_{i,1}, s_{i,2}, ..., s_{i,Q}\},\ s_{i,k} = (x_{i,k}, y_{i,k}, z_{i,k}),\ i = 1, 2, ..., P-1 \tag{3}$$

As it was mentioned before the overall process of calculating the path in 3D space is split on two stages. First one is determination the path in 2D space and second - extension the original path on the 3D space. All further discussion will be made, without the loss of generality, for a single pair of points defined in (2), and denoted hereafter as: m_1, m_2. Of course all the process can be repeated for any pair of the points.

4 Map of the Terrain

The process of planning the collision free path is based on the assumption that the system is provided with the map that contains data about height of each point of the terrain. In this work the map was modelled using simulated sculpture of the earth's surface. It was made using the mixture of Gaussian. Figure 2 shows an exemplary model of the terrain. The next step is to apply a regular grid of the given resolution and put it on the model. If the single cell of the grid has dimensions equal $c_x \times c_y$ and resolution of the grid is $N \times M$ the model will cover the area of $Nc_y \times Mc_x$. With each cell there is related the altitude attribute that is determined as a maximal height in the region covered by given cell. This results in attributing the height parameter to each cell of the grid. Let us define the model of the workspace as the matrix:

$$H = \{h_{i,j}\} \quad i = 1,2,...,N,\ j = 1,2,...,M,\ h_{i,j} \in \Re \tag{4}$$

Such model is the base for the planning algorithm.

5 Planning

5.1 Probabilistic Approach

Since the search space in 3D models is large that implies hard computations have to be performed in order to find the solution. The probabilistic approach allows to simplify the problem. Instead of searching through entire space a number of representative elements are selected. For instance if the map has a resolution 100x100 in the process of searching the solution requires 10000 elements to be processed.

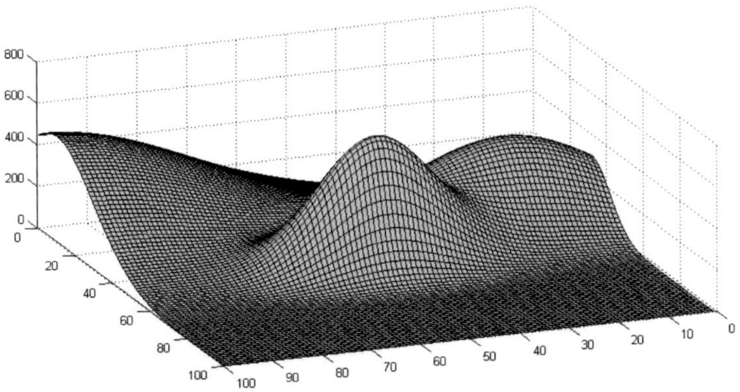

Fig. 2 Simulated model of the terrain

Therefore the number Z of representative elements from the model (4) is randomly selected. The planner searches the optimal solution using the reduced space. Let us denote the randomly selected set of representative samples of the original model as:

$$R = \{r_1, r_2, ..., r_Z\} \tag{5}$$

where

$$r_i = \{k, l\},\ k = \langle 1, N \rangle \text{ and } l = \langle 1, M \rangle \tag{6}$$

In (6) the elements k, l are integer values selected in a random way from the given range. An exemplary containing 25 elements, random realisation of the subspace is presented in fig. 3. Using representative elements contained in (5) the process of search the collision free path is performed. If the size of the (5) is not large enough or the distribution of points does not allow to find admissible solution, the process of selecting the representative subspace is repeated or the size of the set (5) is increased.

5.2 Graph Model

The next stage of the process of finding the solution is creating the model that will be a base for the planning process. Here in this study the graph representation of the random subspace is proposed. Hence the model has a form of weighted graph defined as:

$$W = (V, E) \tag{7}$$

where

$$V = \{v_1, ..., v_{Z+2}\},\ E = \{v_i, v_j\},\ w_{i,j} \neq \infty,\ \ i, j = 1, 2, ..., Z+2 \tag{8}$$

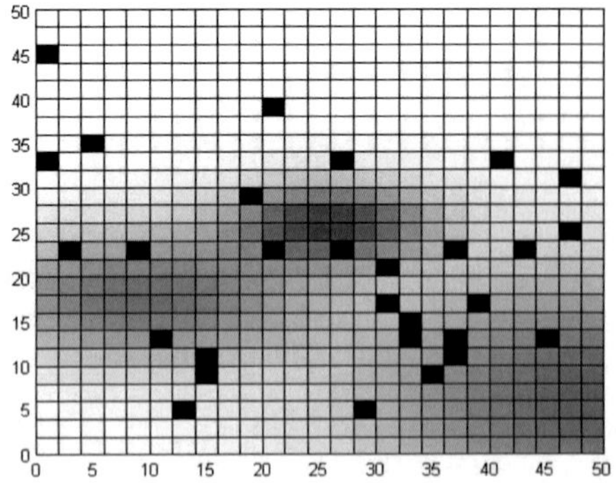

Fig. 3 An exemplary subspace generated in a random way

In (8) v_1 and v_{Z+2} are equal to m_1 and m_2 correspondingly which are defined in (2). This set defines the vertices of the graph (7). Second part of the graph model is the set of edges E. It defines mutual, spatial relations between elements defined by V as well as costs of transitions between any given pair of vertices. Calculation of costs is a crucial part of the model creation. All the process of planning is based on the graph representation so the quality of the model is critical for this process. In this the cost between two elements of the set V is calculated as:

$$w_{i,j} = \left\| v_i - v_j \right\| (1 + S_{i,j}) \tag{9}$$

where $\left\| v_i - v_j \right\|$ denotes euclidean distance between ith and jth vertex, and:

$$S_{i,j} = \sum_{k=1}^{K} L_k \Big|_{v_i}^{v_j} \tag{10}$$

is a sum of altitudes read from the map (4) along the line that connects vertex ith with vertex jth. This mapping is done according the following rule:

$$L_k = \begin{cases} 0 & \text{for altitude} < H_0 \\ H_{m,n} & \text{for altitude} >= H_0 \end{cases} \tag{11}$$

As a result of determining costs of transitions between all pairs of vertices the cost matrix is created. It is denoted as:

$$D = \left\{ S_{i,j} \right\} \; i, j = 1, 2, ..., Z + 2 \tag{12}$$

5.3 Solution

For the representation given by (12) searching the path of minimal cost is performed. There are many effective algorithms for minimal path search [4] but in this study the Dijkstra's algorithm was chosen. Since the graph defined by (12) is the complete one, always exists the path connecting vertices v_1 and v_{Z+2}. As a result of the graph search the path that connects those vertices is found:

$$P^* = \{p_1, ..., p_P\}, \; p_1 = v_1, \; p_P = v_{Z+2}, \; 2 \leq P \leq Z + 2 \tag{13}$$

each element of the set (13) is a point in a 2D space. Next stage consists of transforming this solution into 3D space. This process is done by attributing each point of the (13) with appropriate altitude. This is done in the following way: For each pair p_k, p_l of points from the set (13) it is checked if for any point p_i that lays on the line between p_k, p_l the height determined by (4) is lower than H_0. If yes, the point is attributed with the coordinate $z_i = H_0 + \Delta H$ and this point is added to the solution. As a result of this iterative procedure the final solution defined in 3D space is determined:

$$P_F^* = \{p_{F1}, p_{F2}, ..., p_{FR}\},\ P \leq R \tag{14}$$

where each element p_{Fi} $i = 1, 2, ..., R$ defines a point in 3D space that has coordinates (x_i, y_i, z_i).

6 Simulations

In order to prove the efficiency of the presented method a number of simulation experiments has been performed. Here one relevant experiment is presented as an illustration of the functioning of the method. The map of the terrain that contains information about height was simulated using Gaussian mixtures. The modelled terrain has an altitude from the range 0 to 800 [m]. The map has a resolution 100 × 100. The grid of a resolution 25 × 25 was applied in order to create model of representation. The assumed altitude of the UAV flight was 150 [m]. The initial and the target point of the flight were set to (2,2) and (20,5) correspondingly. The size of the representative, randomly selected space was set to 25. Figure 4 presents results of the planning. This is and exemplary random realisation of the planning process, increasing the number of trials the path can be improved. Also increasing the size of the representative subspace the solution can tend to the optimal one. But due to the fact that the aim of the planner is only to generate way points of the flight trajectory in many cases the solution is feasible and can be applied to effective controlling the UAV.

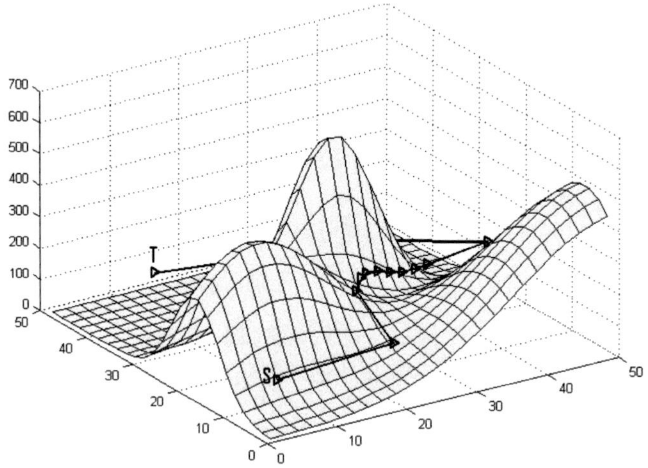

Fig. 4 An exemplary solution of the path planning process in 3D space, obtained obtained using the proposed method

7 Conclusions

In this paper an probabilistic approach to the problem of collision free path planning in 3D space was presented. The target application of the discussed method is determination of the way points of the flight of the UAV. Presented approach allows to find feasible path in 3D space simultaneously reducing the computations. Thanks to the simplification that consists in splitting the process on two stages the method is both effective and simple (in terms of computation costs). The simplicity was the one of the most important assumptions taken for the design of the UAV control system and the presented method satisfy it. Since the method is not deterministic it generates different solutions. After a number of simulations it was checked that all the generated solutions are also feasible ones and they are satisfactory for the overall UAV control system.

References

1. Adolf, F.M., Hirschmuller, H.: Meshing and Simplification of High Resolution Urban Surface Data for UAV Path Planning. Journal of Intelligent and Robotic Systems 61, 169–180 (2011)
2. Barraquand, J., Kavraki, L., Latombe, J.C., Motwani, R., Li, T.Y., Raghavan, P.: A random sampling scheme for path planning. International Journal of Robotics Research 16, 759–774 (1997)
3. Barraquand, J., Langlois, B., Latombe, J.C.: Numerical Potential-Field Techniques for Robot Path Planning. IEEE Transactions on Systems Man and Cybernetics 22, 224–241 (1992)
4. Deo, N.: Graph Theory with Applications to Engineering and Computer Science. Prentice Hall Series in Automatic Computation (1974)
5. Fujimura, K., Samet, H.: A Hierarchical Strategy for Path Planning Among Moving Obstacles. IEEE Transactions on Robotics and Automation 5, 61–69 (1989)
6. Ge, S.S., Cui, Y.J.: New Potential Functions for Mobile Robot Path Planning. IEEE Transactions on Robotics and Automation 16, 615–620 (2000)
7. Hsu, D., Latombe, J.C., Motwani, R.: Path Planning in Expansive configuration spaces. International Journal of Computational Geometry and Applications 9, 495–512 (1999)
8. Hwang, Y.K., Ahuja, N.: A Potential-Field Approach to Path Planning. IEEE Transactions on Robotics and Automation 8, 23–32 (1992)
9. Kambhampati, S., Davis, L.S.: Multiresolution Path Planning for Mobile Robots. IEEE Journal of Robotics and Automation 2, 135–145 (1986)
10. Kant, K., Zucker, S.W.: Toward Efficient Trajectory Planning - the Path-Velocity Decomposition. International Journal of Robotics Research 5, 72–89 (1986)
11. Kavraki, L.E., Kolountzakis, M.N., Latombe, J.C.: Analysis of Probabilistic Roadmaps for Path Planning. IEEE Transactions on Robotics and Automation 14, 166–171 (1998)
12. Kavraki, L.E., Svestka, P., Latombe, J.C., Overmars, M.H.: Probabilistic Roadmaps for Path Planning in High-Dimensional Configuration spaces. IEEE Transactions on Robotics and Automation 12, 566–580 (1996)
13. LaValle, S.M.: Planning Algorithms. Cambridge University Press (2006)
14. Lumelsky, V.J., Mukhopadhyay, S., Sun, K.: Dynamic Path Planning in Sensor-Based Terrain Acquisition. IEEE Transactions on Robotics and Automation 6, 462–472 (1990)

15. Lumelsky, V.J., Stepanov, A.A.: Dynamic Path Planning For a Mobile Automaton With Limited Information on The Environment. IEEE Transactions on Automatic Control 31, 1058–1063 (1986)
16. Lumelsky, V.J., Stepanov, A.A.: Path-Planning Strategies For a Point Mobile Automaton Moving Amidst Unknown Obstacles of Arbitrary Shape. Algorithmica 2, 403–430 (1987)
17. Nakamura, Y., Mukherjee, R.: Nonholonomic Path Planning of Space Robots Via a Bidirectional Approach. IEEE Transactions on Robotics and Automation 7, 500–514 (1991)
18. Nikolos, I.K., Valavanis, K.P., Tsourveloudis, N.C., Kostaras, A.N.: Evolutionary Algorithm Based Offline/Online Path Planner for Uav Navigation. IEEE Transactions on Systems Man and Cybernetics Part B-Cybernetics 33, 898–912 (2003)
19. Paik, D.S., Beaulieu, C.F., Jeffrey, R.B., Rubin, G.D., Napel, S.: Automated Flight Path Planning for Virtual Endoscopy. Medical Physics 25, 629–637 (1998)
20. Pehlivanoglu, Y.V., Baysal, O., Hacioglu, A.: Path Planning for Autonomous Uav Via Vibrational Genetic Algorithm. Aircraft Engineering and Aerospace Technology 79, 352–359 (2007)
21. Thrun, S., Burgard, W., Fox, D.: Probabilistic Robotics. MIT Press (2005)
22. Tisdale, J., Kim, Z., Hedrick, J.K.: Autonomous UAV Path Planning and Estimation An Online Path Planning Framework for Cooperative Search and Localization. IEEE Robotics and Automation Magazine 16, 35–42 (2009)
23. Zhu, D., Latombe, J.C.: New Heuristic Algorithms for Efficient Hierarchical Path Planning. IEEE Transactions on Robotics and Automation 7, 9–26 (1991)

Multi-robot Task Planning Problem with Uncertainty in Game Theoretic Framework

Krzysztof Skrzypczyk and Martin Mellado

Abstract. An efficiency of an multi-robot systems depends on proper coordinating tasks of all robots. This paper presents a game theoretic approach to modelling and solving the pick-up and collection problem. The classical form of this problem is modified in order to introduce the aspect of an uncertainty related to an information about the workspace inside of which robots are intended to perform the task. The process of modelling the problem in game theoretic framework, as well as co-operative solution to the problem is discussed in thise paper. Results of exemplary simulations are presented to prove the suitability of theapproach presented.

Keywords: task planning, uncertainty, game theory, multi-robot.

1 Introduction

In multi-robot systems the primary, complex task is distributed between a number of robots that can perform some simple operations called further sub-tasks. Those systems benefit from the property that even complex process can be solved by a number of simple robotic units. Considering mobile robotics domain, many practical applications of such systems can be pointed out. That is enough to mention a problem of a transportation of large, and heavy parts by a number of smaller transporters, exploration task as well as pick up and collection problem. The results of application of

Krzysztof Skrzypczyk
Silesian University of Technology, Institute of Automatic Control, Akademicka 16, 44-100 Gliwice, Poland
e-mail: krzysztof.skrzypczyk@polsl.pl

Martin Mellado
Instituto de Automatica e Informatica Industrial, Universidad Politecnica de Valencia, Camino de Vera 14, 46022 Valencia, Espana
e-mail: martin@ai2.upv.es

A. Nawrat et al. (Eds.): Advanced Technologies for Intelligent Systems, SCI 440, pp. 69–80.
springerlink.com

multi-robot systems are two fold. On the one hand we benefit from the innate, aforementioned features of these systems like scalability and their distributive nature. But on the other hand, there is a need of coordination of actions of particular units which is not a simple task. Considering distributed mobile robot systems, in many cases the problem of sharing the common workspace and resources appears. This implies that, without proper method of coordination even simple task of moving robots inside of an empty workspace can not be performed. The problem is getting more complex if the information provided to the system is uncertain. In spite of the problems related, advantages of these systems cause they have been intensively investigated. Therefore a series of different methods of solving the problem of coordination have been proposed suggested by different authors [1, 2, 7, 8, 9, 10, 16].

One of the tools which suites well to the discussed problem, is the Game Theory, which provides a smart framework for modelling and solving the problems that have conflicting nature. It is easy to notice that in multi-robot systems, mutual interactions between individual robots may result in conflict, which is often caused by the fact of sharing common resources. That is the reason many researches try to apply this framework to model different aspects of multi-robot systems [3, 4, 5, 11, 15]. In this paper an approach to model and solve the pick up and collection task in multi robot environment is presented. This problem is a popular benchmark in Multi-Agents Systems which allows to verify different control, coordination and task distribution strategies. Classical formulation of this problem, which has been investigated in many papers [12, 13, 14] does not take into account uncertainty of information about the workspace the team of robots is provided with [10]. To extend of the primary pick-up collection problem it is assumed that the number as well as location of the objects the team is intended to collect is uncertain. In this case the problem is getting more interesting, and hard to solve by off-line task planners. Such modification of the problem is the point of this paper. The rest of the paper is organized as follows. After introducing the basic control system framework a model of the workspace is described. Next part of the article define the tasks particular robots are able to perform. After this the problem statement is presented. Another part of the article describes the process of modelling the problem in the game theoretic framework. In the last section simulations that prove the appropriateness of presented approach are presented and discussed.

1.1 The System Framework Overview

In this section a brief overview of the system's framework is presented. Figure 1 illustrates conceptual architecture of the multi-robot control system. It is assumed that the system is the centralized one. That means that the control is determined by the module responsible for the task planning and then sent to the individual robotic units. The task planner computes the set of actions that are to be executed by each robot. That implies the system works in a synchronous way. The problem solver cannot send the new set of actions to the group of robots, until the last robot reports completion of its task. Therefore the system should not be perceived as a distributed control system (multi-agent). The control is centralized and sent to the individual robots synchronously. On the other hand the information about the workspace is collected and then updated in a distributive way. The model of the workspace introduced to the system in the beginning of the process, is successively updated by individual robots. In the system described, the blackboard exchange mechanism is applied, what means that each robot has an access to some system variables, and is able to read them and overwrite them, while collecting new information.

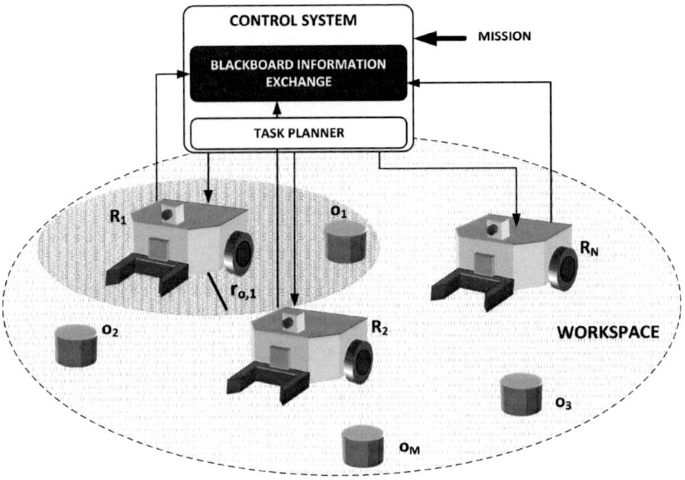

Fig. 1 Conceptual diagram of the system

2 Model of the Workspace

The pick-up and collection problem consists of collecting by the group of robots a number of objects scattered inside the workspace. The system is provided with information about the location of objects described in the Cartesian coordinates system fixed to the workspace. However, the information about the fact that there is the given object on the known position is not certain one. This uncertainty must be taken into account during the process of planning the task, what implies it must be included in the model of the workspace. Moreover each robotic unit is assumed to have some operational range, that describe its possibility of the movement during the single cycle of the process. This aspect of the limited mobility of robotic units must also be included in the model. Taking aforementioned aspects into account, the workspace is modeled as a weighted visibility graph, which determines possible movements of individual units in the given state of the process. The graph is defined as $W = (V, E)$ where $V = \{v_1, v_2, ..., v_M\}$ is the set of vertices of the graph, that represent M objects deployed inside of the workspace. The following features are related to the given vertex:

$$v_i = [x_{o,i}, y_{o,i}, P_{o,i}, r_{o,i}] \tag{1}$$

In (1) variables $(x_{o,i}, y_{o,i})$ denote the position of ith object, $P_{o,i}$ is the probability of the fact that ith object is located inside of the circle of the centre fixed in $(x_{o,i}, y_{o,i})$ and the radius $r_{o,i}$. Since for considered problem both shape and orientation of objects is not important, each object is described only by position of geometrical centre. Second part of the model is the set of edges E that describes spatial relations between objects. As it was mentioned before, each robot is assumed to have a given operational range $r_{R,i}$ for $i = 1, 2, ..., N$, where denotes a number of robots. That means it can only reach an object that is closer than the range of the robot. Moreover, a robot cannot move directly to an object if there is another object on the way. The reason of such assumption is that such movement can disturb the searching or picking-up process proceeded by other robot. The set of objects that lay on the the way between ith and jth object is denoted by:

$$V_{i,j}^{obst} = \{v_k, ..., v_l, ..., v_m\} \subset V \tag{2}$$

Thus the edges of the graph are defined:

$$E = \{v_i, v_j, w_{i,j} : v_i, v_j \in V, w_{i,j} \in \mathbf{R}\} \tag{3}$$

The weighting factor $w_{i,j}$ in (3) is calculated using the formula:

$$w_{i,j} = \begin{cases} \left\|v_i - v_j\right\| & \text{if} \left\|v_i - v_j\right\| < r_{R,i} \wedge V_{i,j}^{obst} = \phi \\ \infty & \text{otherwise} \end{cases} \tag{4}$$

The weighting factor is equal to the distance between the object and the robot if this distance is smaller than operational range of the robot. Also must be assured that there is no any other object laying between the robot and the given object. Otherwise the weighting factor is equal to infinity. The model contains information about the workspace the system is provided with. The real state of the workspace is described by:

$$V^{real} = \{P_{r,1}, P_{r,2}, ..., P_{r,M}\} \;\; P_{r,i} = 0, 1 \tag{5}$$

Elements of this set describe the fact if really there is the given object inside of this workspace or not. In fact the set (5) denotes a true, initial state of the workspace and in general can differ from the description given by (1).

2.1 Robots and Their Actions

Within the workspace W operate N mobile robots. The group of robots is denoted as a set $R = \{R_1, R_2, ..., R_N\}$. The state of ith robot is defined by its position and described by a vector:

$$X_i = [x_{R,i}, y_{R,i}] \tag{6}$$

where elements of the (6) denote the position of ith robot described in the coordinate frame of the workspace W. Since it is not important for considered problem, the orientation of the robot is neglected. In terms of the assumed model of the workspace as well as the process of task planning, the robot can be located only in finite number of places, which are defined by objects' location. Therefore, motion between these locations is assumed to be solved by a path planner avoiding any possible collision. Thus the position of the ith robot, in terms of the model, is defined

$$m_i = m,\ m = 1,2,...,M \ \text{for} \ \sqrt{(x_{R,i} - x_{o,m_i})^2 + (y_{R,i} - y_{o,m_i})^2} < r_{o,m_i} \tag{7}$$

Therefore the state of the team in the given discrete moment n, $n = 1,2,...$ of the process is given by:

$$M(n) = \{m_i\},\ i = 1,2,...,N \tag{8}$$

Each robot depending on its state (7) is able to perform a given number of actions. The set of actions admissible in the given state m_i for ith robot is denoted by:

$$A_i^{m_i} = \{a_{i,1}, a_{i,2}, ..., a_{i,K_i}\} \tag{9}$$

where K_i is a number of admissible actions of the ith robot in the state m_i. Each action the given robot can take is the one of the three following ones.

- $GoTo(R_i, v_j)$ - is the action that consists in moving the ith robot close to the jth object. This action, in order to be admissible, must satisfy the precondition: $w_{m_i,j} \neq \infty$, whereas a result of applying this action is described by the postcondition: $m_i = j,\ P_{o,j} = P_{r,j}$

- $PickUp(R_i, v_j)$ - is second action that can be taken by a given robot. It consists in picking up and collecting the object v_j by the robot R_i. This action can be performed if it is satisfied the following: $m_i = j,\ P_{o,j} \neq 0\ v_{m_i} \notin V^{col}$ where V^{col} denotes the set of objects that have already been collected. Applying this action results in the postcondition: $P_{o,j} = 0;\ \ V^{col} = V^{col} \cup v_j$

- $Wait(R_i)$ - this is third action the given robot can take. The result of applying this action is stopping the robot for one planning cycle. This action does not need any preconditions.

2.2 Problem Statement

Using the introduced notation the problem can be clearly sated as follows. For each descrete moment of time n which is the planning cycle, find a set of actions $A_0(n) = \{a_{0i}\}$ $i = 1,2,...N$, $n = 1,2,...$ that if applied will results in satisfying the following:

$$V = V^{col} \tag{10}$$

It is required the condition (10) to be satisfied, after as small as possible number of planning cycles. Moreover the aim of the planning process is an uniform distribution of collected objects that reduce the amount of energy spent by the team. The (10) is a condition of terminating the collection process and means that all known objects are collected. It is important to stress that all the process of task planning is not the off-line one. Since there is an uncertainty of the information, the tasks are scheduled in the step by step mode, while the information is being updated by the robots performing the mission. That is the reason that the result of the task planning might not be optimal. It depends strongly on the level of uncertainty of information the system is provided with.

3 Modelling the Problem

The pick-up and collection problem is a typical example of a process when multiple robotic units must share common resources during execution the common mission. That implies mutual interactions of the units can lead to ineffective or even to inability of execution of the mission. In this section the process of modelling the pick-up and collection problem in game theoretic framework is described. The process of execution of the considered task can be perceived as a game between individual robotic units (players). Each of them is able to perform a set of the actions described in section 2.1. Depending on the given combination of actions taken by individuals and the current state of the process, the game results in some cost the team is trying to minimise (in case of team problem). Let us model considered process as a game related to the given discrete moment of the planning process: $G(n) = \{N, A(n), I(n)\}$, where where N denotes number of players (robots) taking part in the game, $A(n)$ is an action space of the given nth step of the process:

$$A(n) = A_1^{m_1} \times A_2^{m_2} \times ... \times A_N^{m_N} \tag{11}$$

where $A_i^{m_i}$ is an action set of the ith robot in the state m_i defined by (9). Third component of the game is the cost function I. The game theory derives from the economical sciences and in this context the cost function denoted the amount of costs that the players had to pay for taking individual combination of strategies. In terms of robotics, the cost function has no physical interpretation, and the value the function returns is considered as a numerical quantity that is to be minimized. Designing the cost function is the key point in creating the game theoretic framework for the considered problem. The value of the cost function, arguments of which are particular actions of the robotic units, reflects the quality of execution of the considered task. The form of this function must encapsulate all the features of multi robot execution of the mission. In discussed approach the following form of the cost function was applied:

$$I = \sum_{i=1}^{N} I_i(a_k), a_k \in A_k^{m_k} \; k = 1,2,...,N \tag{12}$$

Each component I_i is the cost function related to the ith robot. This component of the cost functions, consists of the two parts:

$$I_i(a_k) = I_{exp,i} + I_{rew,i} \tag{13}$$

The first one is related to profitability of the action applied. Since a presence of the object in the location is given with some probability, there is a need to estimate the costs of moving the robot to the given position. The higher probability of the presence of the object and the lower is the distance to it, the more profitable is taking the considered action. This component for the action $a_k = GoTo(R_i, v_j)$ is defined by:

$$I_{\exp,i}(a_k) = \begin{cases} \infty & \text{if} \quad v_j = v_i, \; j,j = 1,...,N \\ \alpha \frac{P_{o,a_i}}{L_{i,v_i}} + \beta \frac{P^*_{o,v_i}}{L^*_{o,v_i}} + b & \text{otherwise} \end{cases}, \quad k = 1,2,...,N \tag{14}$$

where P_{o,a_i} is the probability of existence of the object v_i selected by ith robot as a result of action a_k. The value L_{i,v_i} is the cost of moving ith robot to the given object a_i. This value is equal to the weighting factor w. The value of the other part of described component is related to proximity of the object v_i, to the closest uncollected object. Thus P^*_{o,v_i} and L^*_{o,v_i} denote the probability of existance of the object closest to v_i and distance between the closest, uncollected object and the object v_i. The role of coefficients α, β as well as the bias b is adjusting the model. The value of this component for actions $PickUp$, and $Wait$ is equal to 0.

The value given by second component in (13) can be described as a reward taken by individual player, for picking up the given object. This value depends on the state of the overall process, and can vary depending on the distribution of number of objects collected by individual robots. The aim is to provide uniform distribution and obtain the best possible value of the cost function. Therefore the value of this factor, defined for $PickUp$ action is given by:

$$I_{rew,i}(a_k) = \begin{cases} -R & \text{if } \hat{\delta}_i > 0 \wedge m_{v0} = v_i \wedge \hat{L}^*_i < L_{v_k,v_{k0}} \\ R & \text{otherwise} \end{cases} \tag{15}$$

where R is some positive, real value. Therefore, this factor can be positive or negative according to whether the decision taken by ith robot is cooperative or noncooperative The action is considered to be noncooperative one if the conditions pointed in (15) are satisfied. The first part of the condition determine if number of collected objects by ith robot is not lowest one (including the object collected as a result of action a_k). The value is determined from:

$$\hat{\delta}_i = \hat{N}_i - \min_{k=1,...,N,k\neq i} (\hat{N}_k) \tag{16}$$

where $\hat{N}_k, \quad k = 1,...,N$ is the number of collected objects by kth robot, after taking the action a_k. Next conditions determine if the object v_i the ith robot is about to pick up is the closest uncollected object for other robot. To know if it is, it is checked if the distance $\hat{L}_i^*$ to other closest object for ith robot is smaller than $L_{v_k,v_{k0}}$. Such form of the component stimulates the cooperative behaviour of the team. It allows to provide uniform distribution of number of collected elements as well as minimise the energy spent by a team. This term is equal to 0 for *GoTo* and *Wait* actions.

4 Solution

The problem stated in this paper is called in terms of game theory the team problem which consists in optimisation the common cost function. Moreover the considered problem is cooperative one with total exchange of information and successive synchronisation of task execution of individual robot. It is important to stress again that the task execution process is the iterative one with successive update of knowledge made by individual team mates. Therefore the solution of the single nth step of the process is given by:

$$A_0(n) = \{a_{10}, a_{20}, ..., a_{N0}\} = \min_{a_1,...,a_N} I \tag{17}$$

Applying the solution results in execution of the part of the primary problem and update the knowledge about the workspace. The solution (17) is applied until the stop condition is satisfied. The stop condition is defined by equality of the sets $V = V^{col}$ what means that all the objects have been collected. The process also can stop when all the actions determined by (17) are *Wait* actions. This second situation may occur when there are objects described in model by very low probability factor. In this case if possible benifits from checking up the existance of the objects are lower than energy spent for this action the process may by stopped.

5 Simulation Results

In order to verify our approach the afore-described methodology was implemented in the MATLAB environment. A number of simulations have been performed to evaluate the proposed methodology of modelling and solving the pick up collection task. In this section only two, selected and relevant experiments are presented and discussed. In all presented simulations two robots $\{R_1, R_2\}$ are intended to collect a set $\{v_1, v_2, ..., v_M\}$ of objects where the number of objects is equal to 10. Inside the workspace, there are M objects distributed in a random way. The probabilities of existence are randomly fixed to the given objects. In all (excluding the first one) experiments 30% of objects are described by low probability factor. Among these objects there are objects that do not exist in the workspace. Thus difference between the real world and the model of the process is modeled. In first experiment the discussed method was verified using different scenario. This time ten objects were randomly displaced inside of the workspace. Existence of each object was described by probability factor (depicted by number outside the circle). Initial locations of robots are close to each other what implies some possible difficulties in task allocation. There are 2 objects described by low probability factor in the model, and which did not exist in reality (objects depicted with double circle). Modeled scenario is presented in fig. 2 while fig. 3 presents the solution of the problem, which is the sequence of robots actions. It is worth to notice that also in this scenario, the task was performed in a cooperative way. In stage 7 first robot leaves the object instead of picking it up and moves to 6th object. That allowed to obtain uniform distribution of collected objects between the robots.

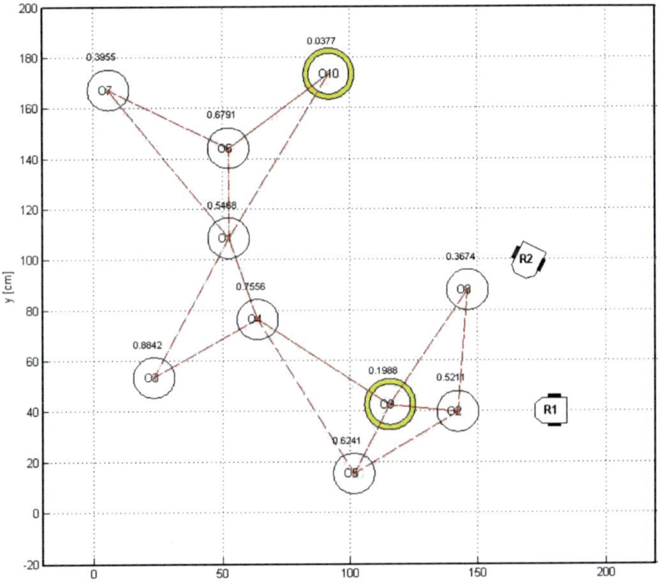

Fig. 2 The layout of the objects and the model of the workspace used in first experiment

The goal of the next experiment is showing the functioning of the method in the case when the objects are split into two separated subgroups (fig.4). Also the robots start the work from two distant locations. In this experiment three objects were modelled as not existing ones, described by low probability factors. Analysing the

Process stage no	ROBOT 1	ROBOT 2
1	GoTo(R1,o2)	GoTo(R2,O8)
2	PickUp(R1,o2)	PickUp(R2,o8)
3	GoTo(R1,o5)	GoTo(R2,O9)
4	PickUp(R1,o5)	GoTo(R2,o4)
5	Wait(R1)	PickUp(R2,O4)
6	GoTo(R1,o4)	GoTo(R2,o3)
7	GoTo(R1,O1)	PickUp(R2,03)
8	GoTo(R1,o6)	GoTo(R2,O1)
9	PickUp(R1,o6)	PickUp(R2,o1)
10	GoTo(R1,o7)	GoTo(R2,o10)
11	PickUp(R1,o7)	Wait(R2)

time

Fig. 3 Sequence of robot actions - result of execution of the task stated in first experiment

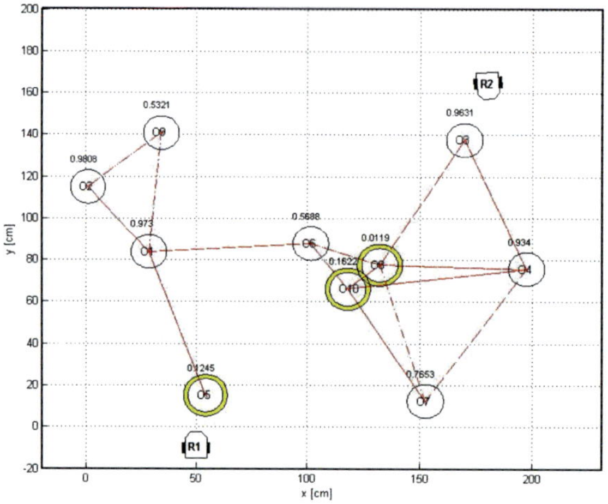

Fig. 4 The layout of the objects and the model of the workspace used in second experiment

solution (fig. 5) it is easy to notice that the behaviour of the team is quite reasonable and intuitive in terms of human behaviour. Robot 1 after picking-up 9th object, waits and let the other robot collect the rest of objects. Thus energy saving aspect of the team behaviour can be noticed.

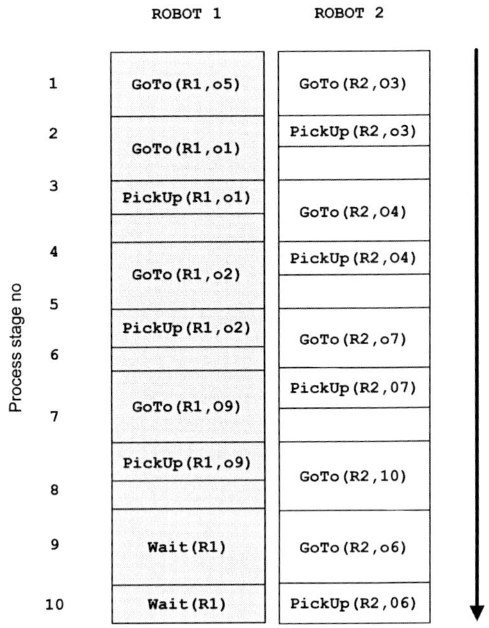

Fig. 5 Sequence of robot actions - result of execution of the task stated in second experiment

6 Conclusion

In this paper an approach to model and solve the pick up and collection task in multi-robot environment was presented. The discussed problem is a popular benchmark in Multi-Agents Systems which allows to verify different control, coordination and task distribution strategies. The extended version of the problem, that consists in introducing uncertainty of the model was the key point of this article. It was assumed that the number as well as location of the objects the team is intended to collect is uncertain. A game theoretic approach to model and solve the aforementioned problem was applied. Because the knowledge of the system about environment is uncertain it is difficult to apply off-line planning. Instead of this an iterative step-by-step execution of the task was applied. Implications of such approach are two fold. On one hand it is possible to acquire knowledge about the environment and the real state of the process. On the other hand this approach can not guarantee the optimal execution of the task. Nevertheless a number of simulation, three examples of which are presented in the paper, confirm the appropriateness of the approach. Results of the simulations show that the method behaves in a rational (in terms of human behaviour) way and the obtained solutions are acceptable. Such good results are obtained of course in case of providing step by step synchronisation of the process. Future investigation will be focused on examining the problem when actions are not synchronised. In this case another uncertainty factor related to the effect of the actions taken by robots, should have to be considered.

References

1. Alami, R., et al.: Toward human-aware robot task planning. In: Proc. of AAAI Spring Symposium, Stanford (USA), pp. 39–46 (2006)
2. Baioletti, M., Marcugini, S., Milani, A.: Task Planning and Partial Order Planning: A Domain Transformation Approach. In: Steel, S. (ed.) ECP 1997. LNCS, vol. 1348. Springer, Heidelberg (1997)
3. Desouky, S.F., Schwartz, H.M.: Self-learning Fuzzy logic controllers for pursuit-evasion differential games. Robotics and Autonomous Systems (2010), doi:10.1016/j.robot.2010.09.006
4. Harmati, I., Skrzypczyk, K.: Robot team coordination for target tracking using fuzzy logic controller in game theoretic framework. Robotics and Autonomous Systems 57(1) (2009)
5. Kaminka, G.A., Erusalimchik, D., Kraus, S.: Adaptive Multi-Robot Coordination: A Game-Theoretic Perspective. In: Proc. of IEEE International Conference on Robotics and Automation, Anchorage, Alaska, USA (2002)
6. Kok, J.R., Spaan, M.T.J., Vlassis, N.: Non-communicative multi-robot coordination in dynamic environments. Robotics and Autonomous Systems 50(2-3), 99–114 (2005)
7. Klusch, M., Gerber, A.: Dynamic coalition formation among rational agents. IEEE Intelligent Systems 17(3), 42–47 (2002)
8. Kraus, S., Winkfeld, J., Zlotkin, G.: Multiagent negotiation under time constraints. Artificial Intelligence 75, 297–345 (1995)
9. Kraus, S.: Negotiation and cooperation in multiagent environments. Artificial Intelligence 94(1-2), 79–98 (1997)
10. Mataric, M., Sukhatme, G., Ostergaard, E.: Multi-Robot Task Allocation in Uncertain Environments. Autonomous Robots (14), 255–263 (2003)
11. Meng, Y.: Multi-Robot Searching using Game-Theory Based Approach. International Journal of Advanced Robotic Systems 5(4) (2008)
12. Jones, C., Mataric, M.: Adaptive Division of Labor in Large-Scale Minimalist Multi-Robot Systems. In: Proc. of IEEE/RSJ International Conference on Intelligent Robots and Systems, Las Vegas, USA, pp. 1969–1974 (2003)
13. Sariel, S., Balch, T., Erdogan, N.: Incremental Multi-Robot Task Selection for Resource Constrained and Interrelated Tasks. In: Proc. of 2007 IEEE/RSJ International Conference on Intelligent Robots and Systems, San Diego, CA, USA (2007)
14. Schneider-Fontan, M., Mataric, M.J.: Territorial Multi-Robot Task Division. IEEE Transactions on Robotics and Automation 14(5), 815–822 (1998)
15. Song, M., Gu, G., Zhang, R., Wang, X.: A method of multi-robot formation with the least total cost. International Journal of Information and System Science 1(3-4), 364–371 (2005)
16. Cheng, X., Shen, J., Liu, H., Gu, G.-c.: Multi-robot Cooperation Based on Hierarchical Reinforcement Learning. In: Shi, Y., van Albada, G.D., Dongarra, J., Sloot, P.M.A. (eds.) ICCS 2007. LNCS, vol. 4489, pp. 90–97. Springer, Heidelberg (2007)

The Concept of Collision-Free Path Planning of UAV Objects

Artur Babiarz and Krzysztof Jaskot

Abstract. The paper presents the concept of path planning of unmanned flight the helicopter type of flying object. The presented method is based on probabilistic robot motion planning method called B-PPT. Method B-PPT is built upon the known methods of PRM (Probabilistic Roadmap) and RRT (Rapidly-exploring Random Trees). Using the method of B-PPT can plan the path of mobile robots and manipulation. Planning the path of a flying object was carried out in Cartesian space, taking into account only the position and altitude. The speed of the method depends on two input variables, ie the configuration space discretization step and set number of randomly generated configurations. Effect of size on the results obtained was presented in this work.

Keywords: UAV object, path planning, B-PPT method.

1 Introduction

In recent years, it can be seen very strong interest in collision-free motion planning of robots of various types: stationary, wheeled, biped and flying [15, 16]. This is due to increasing automation and robotics, not only industry but also in everyday life. Robots currently found in almost every area of life from basic household tasks, by medicine, industry, and ending with the tasks in space. On the other hand, seeks to create a robot manipulative, which could fully simulate the movements and behavior [11]. In all these cases the main problem is intended to move all or part of its initial position to final position. To solve this task in question are used, inter alia, probabilistic methods.

Artur Babiarz · Krzysztof Jaskot
Silesian University of Technology, Institute of Automatic Control,
Akademicka 16, 44-101 Gliwice, Poland
e-mail: {artur.babiarz,krzysztof.jaskot}@polsl.pl

A. Nawrat et al. (Eds.): Advanced Technologies for Intelligent Systems, SCI 440, pp. 81–94.
springerlink.com

1.1 PRM Method

PRM method consists of two main phases [5, 7, 8]:

- the pre-processing phase,
- the phase of the inquiry.

During pre-processing is carried out random generation of locations in the obstacle-free space and create a graph G. In general, the whole environment is marked C, in which there are obstacles to occupy the O. Then the obstacle free space is defined as $F = \frac{C}{O}$ [5]. The number generated in this way is very high positions in the order of several thousand. Arrangement of neighboring points relative to the vertex x is dictated by the criterion [1, 5, 7, 8]:

$$N_x = \{y \in V : D(x,y) \leq maxdistance\} \tag{1}$$

After completion of the initial phase of processing phase starts queries, which is looking for a road linking the starting point P_s to the end point P_k. During the phase of inquiry be verified if you can connect these two points with vertices graph G.

1.2 RRT Method

An alternative to the PRM method has become a way of planning movement called the Rapidly - exploring random Teres. LaValle is the originator of that in [10, 12, 13] presented the first principles and algorithm of the method. RRT method is to intend of removing the disadvantages of PRM, which include primarily the problems of narrow passages in the planning and generated a very large number of points in space to obtain a satisfactory result [6]. Furthermore, due to the use of two-phase algorithm was accused of PRM method can not be applied in the planning of on - line (due to time-consuming graph search performed during the pre-processing). Therefore, the RRT method it is intended to generate additional points to keep checking the condition:

$$\rho(x_i, x_{i+1}) \leq \varepsilon \tag{2}$$

where:
ρ - metrics,
x_i, x_{i+1} - successively generated random space points in time i and i +1,
ε - very small positive number.

In the basic RRT method is generating random, from a starting point, successive vertices in obstacle-free space. Used space is called the space of states. For this purpose, a state equation $\dot{x} = f(x,u)$, where x is the state of the system, and u in the entry system. Allowable set of inputs ensures minimizing the distance between the current state x_i and another one x_{i+1} by using the metrics described by the equation (2). This way of the planning movements can be used for holonomic and nonholonomic robots.

2 B-PPT Method

The presented probabilistic methods PRM i RRT and their modifications have one thing in common. All are based on the so-called local scheduler, which consists of connecting the neighboring vertices, which are the points of the space, straight lines. Assuming that an object can only move in straight lines to be rather simplistic (the exception is the work of [14], which uses the functions of class C^1). In addition, a description of the motion as a linear function eliminates the possibility to obtain velocity and acceleration of the object by calculating the derivatives of these functions at points of connection lines. A major problem may also designate an area which is free and occupied by potential obstacles workspace. For manipulation objects process area is often very complicated and difficult to describe with the mathematical formulas. The name of the method of B-PPT is derived from the abbreviation of Probabilistic trajectory planning using B - splines curves. The proposed method is designed mainly for the manipulators of industrial robots. It is based on task forward kinematics and planning in natural coordinates. The genesis of the method B-PTT can be analysis of how the PRM and RRT, which consequently led to the use of certain features of both methods and connections together with algorithms design curves B - splines.

Algorithm B-PPT method can be presented in the following steps:

- Step 1: load vectors of the initial and final position,
- Step 2: detection collisions for the initial and final position, when a collision is detected, go to Step 9,
- Step 3: generate random vectors L setpoints configuration determining the robot,
- Step 4: The collision detection for L vectors of the position of step 3,
- Step 5: Select the location of vectors that meet the selection criterion,
- Step 6: calculation of the B-splines curve,
- Step 7: The collision detection position vectors of coordinates corresponding to the step 6, when a collision is detected go to step 8, otherwise go to Step 9,
- Step 8: remove the checkpoints of the curve and return to step 6,
- Step 9: The end.

The calculation result obtained by changing the coordinates described by polynomials of any degree, depending on the demands made by the user. Moreover, this method is quick and easy to use because of the effective methods of generation of B - splines curves and the advantages of probabilistic methods. Description and a detailed description of the method of B-PPT can be found in [3].

2.1 Selection Criterion

Selection of successive configurations is based on the distance between a given, and the new location in the workspace. They can be described as a selection criterion by formulas [2]:

$$N_c = \{\tilde{c} \in N \mid D(c,\tilde{c}) = mindist\} \tag{3}$$

$$N_c = \{\{\tilde{c_1},\tilde{c_2}\} \in N \mid D(c_k,\tilde{c_1}) < D(c_k,\tilde{c_2})\} \tag{4}$$

where:
c_k - final position,
$\tilde{c},\tilde{c_1},\tilde{c_2}$ - tested at a time position,
$\{\tilde{c_i},\tilde{c_j}\}$ - selected two neighboring positions, i,j=1, 2, ..., m,
m - number of selected positions,
N_c - a set of selected points,
N - the set of all randomly generated points,
D - value of the Euclidean metric.

2.2 B-Spline Curves

Used in the experiments simulated curves are called open uniform B-Spline curves of third degree [9]. B-spline functions given by the formula:

$$B_i^3(t) = \begin{cases} \frac{(2+t)^3}{6} & \text{if } -2 \le t < -1; \\ \frac{(4-6t^2-3t^3)}{6} & \text{if } -1 < t \le 0 \\ \frac{(4-6t^2+3t^3)}{6} & \text{if } 0 < t \le 1 \\ \frac{(2-t)^3}{6} & \text{if } 1 \le t < 2; \\ 0 & \text{if } t \ge 2; \end{cases} \tag{5}$$

While the B-spline curves can be described using the equation:

$$s(t) = \sum_{i=0}^{N-n-1} d_i B_i^n(t) \tag{6}$$

where:
N - number of checkpoints,
n - the degree of function of B-spline,
D_i - i-th control point of the curve.
Rewriting the above equation for the coordinate position is obtained:

$$q^i = \sum_{j=0}^{m} p^j B_j^k(u) \tag{7}$$

where:
m - the number of checkpoints,
k - the degree of B-spline functions,

q^i - i-th coordinate of the position, i = 1-n
n - number of the coordinate position of the robot,
u - fixed increasing sequence of numbers. p^j - drawn at random in the j-th iteration of the coordinate position, which is the control point B-spline curve, satisfying the condition $q^i_{min} \leq p^j \leq q^i_{max}$

In order to ensure the interpolation values of the initial and final coordinates of the position shall be a multiple of the value that is equal to the degree of B-spline function Fig. 1.

Fig. 1 Example of B-spline curve

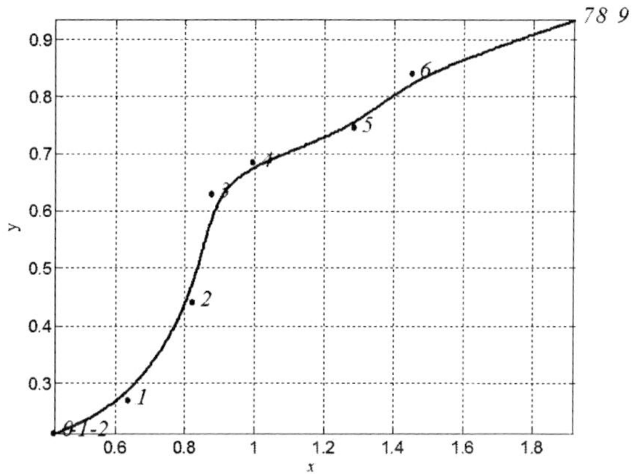

3 UAV Mathematical Model

The proposed flight path generation method is carried out on several assumptions:

1. flight takes place at a constant height,
2. flight control system has a photo area where the road is generated, eg from a digital camera,
3. the control system is supplied only item in the i-th moment (x_i, y_i),
4. path is represented by spline curves, degree 3, which allows you to calculate air speed along the path, which is a preset speed, according to the formula (10).
5. input data are vectors of initial position, final altitude and the area representing obstacles and free space.

Determined path is supplied to the control system (exactly the autopilot), shown in the Fig. 2. The control system based on any given location is calculated in the appropriate control signals (in this case the angles responsible for the flight UAVs). Vector appropriate setting angle is applied to the controller, which using information from sensors maintains commanded the flight parameters. In addition, assume that the UAV is equipped with an autopilot and a digital camera and high speed allow communication with the base station. Due to the required relatively high speed of calculation also assumes that the mathematical operations are carried out on computing unit located in the base station.

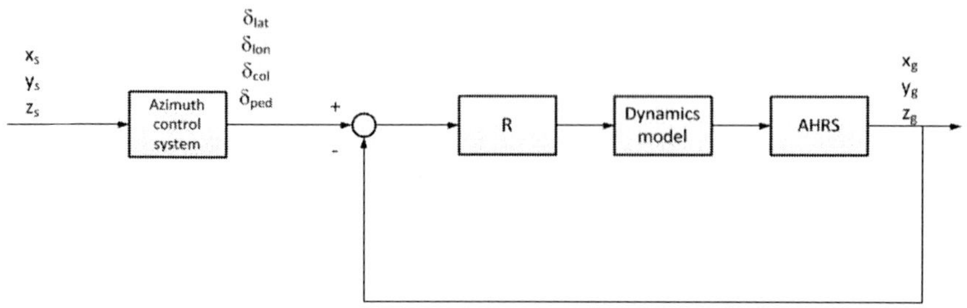

Fig. 2 The system control, where: (x_s, y_s, z_s) - start position, (x_g, y_g, z_g) - goal position, R - regulator, AHRS - Attitude and Heading Reference System, δ_{lat} - aileron servo input, δ_{lon} - ellevator servo input, δ_{col} - collective pitch servo input, δ_{ped} - rudder servo input

Typical UAV dynamics model consists of 12 variables developed in [18]. This model is very accurate, which can cause problems during generation of flight paths. Therefore, in order to facilitate the calculations used a model with six variables (when the UAV has a preset position, altitude and speed):

$$\begin{aligned} \dot{z}_x &= v\cos(\psi) \\ \dot{z}_y &= v\sin(\psi) \\ \dot{\psi} &= \alpha_\psi(\psi^c - \psi) \\ \dot{v} &= \alpha_v(v^c - v) \\ \ddot{h} &= -\alpha_{\dot{h}}\dot{h} + \alpha_h(h^c - h) \end{aligned} \tag{8}$$

where: ψ^c,v^c and h^c are specified course, speed and height passed to the autopilot and α_* are positive constants [4]. This model captures the fundamental dynamics of the system [4, 17]. Further simplification can be assumed that the road will be flown at a constant speed (v^c) and a constant height. UAV dynamics simplifies to:

$$\begin{aligned} \dot{z}_x &= v\cos(\psi) \\ \dot{z}_y &= v\sin(\psi) \\ \dot{\psi} &= \alpha_\psi(\psi^c - \psi) \end{aligned} \tag{9}$$

where: v^c means a constant speed along the path, and ψ^c is the position is sent to the autopilot. Using the dynamics described by equation 9, allows the generator to provide a path similar form. Specifically, the path generator is:

$$\begin{aligned} \dot{x}_r &= v^c\cos(\psi_r) \\ \dot{y}_r &= v^c\sin(\psi_r) \\ \dot{\psi}_r &= u \end{aligned} \tag{10}$$

where: (x_r, y_r) is the current position, ψ_r is a course, v^c is the set linear velocity along the path, and u is a given course.

4 Examples

At this point, will be presented examples of two cases of generation of UAV flight path. They differ in the placement of obstacles in the helicopter flight. In addition, assume that the UAV is equipped with an autopilot and a digital camera and high speed allow communication with the base station. Due to the required relatively high speed of calculation also assumes that the mathematical operations are executed on computing unit located in the base station. Examples used in view of a specific area of the city. Simulations performed by changing the discretization step-flight area and the number of randomly generated vectors of position in space flight. The dimensions given on the graphs are symbolic and are not actual size of the area in which you want to move the UAV. Obstacles are represented as rectangles. It allows to easily check whether a given point lies in the flight path area free from obstacles. The figure 3 shows the first example of the area used in simulations. Simulation used to approximate the objects on the photograph in the form of cuboids, which facilitates collision detection issue.

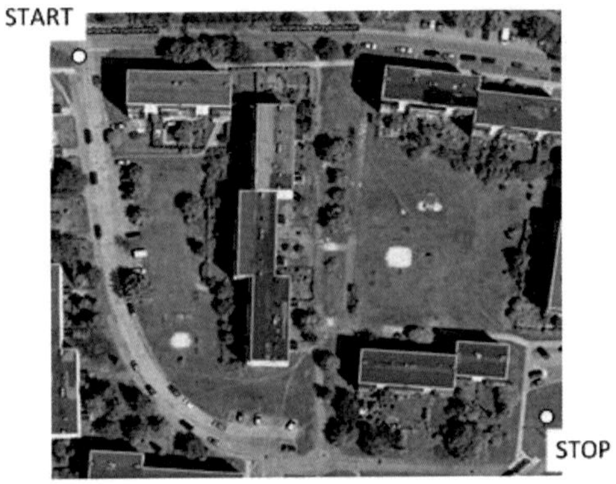

Fig. 3 The first area

In the first example, set the following parameters:

- Number of vectors generated position 200,
- Discretization step 5 [m].

Figure 4 shows the results obtained from simulation.

Graphs represent the flight path of the generated vectors and the distribution of positions used to plan the flight. The result was that in order to solve the problem, the algorithm needed 159 positions.

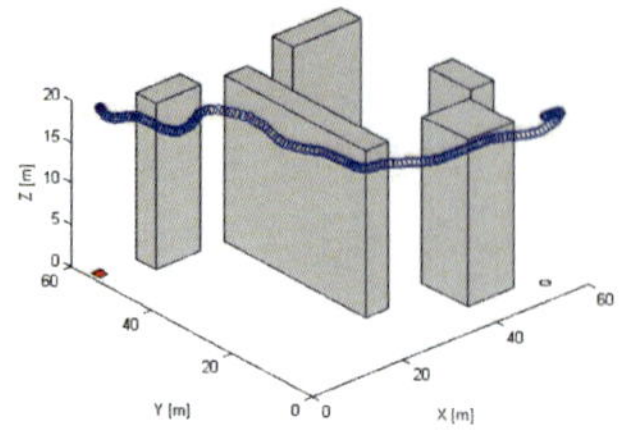

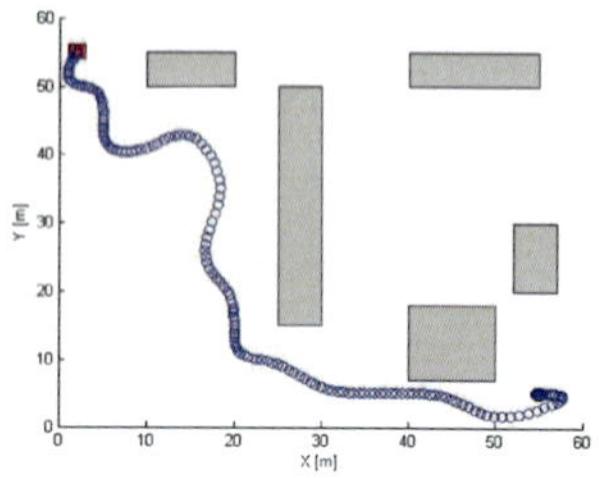

Fig. 4 The simulation: number of position 200, discretization step 5 [m]

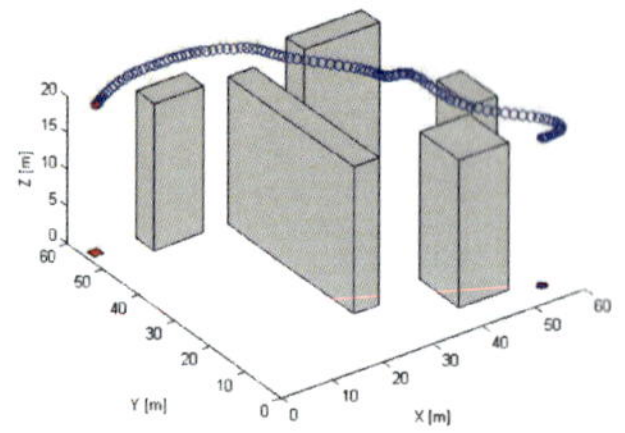

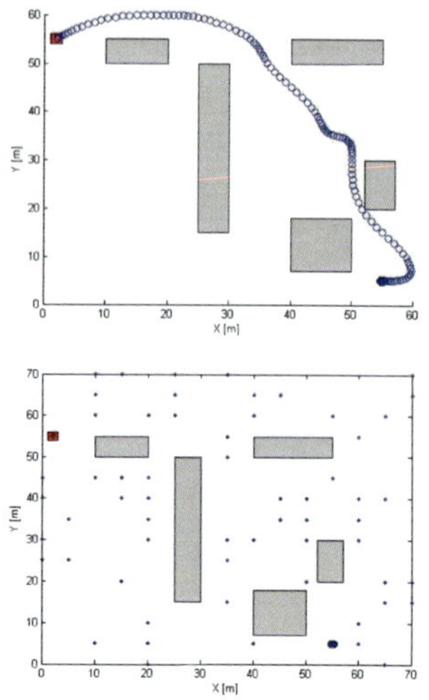

Fig. 5 The simulation: number of position 100, discretization step 5 [m]

In the next simulation step is digitizing the same, but reduced the number of generated positions to 100. Algorithm without any major problems generated collision-free path and the need for this less than half of the vectors of position, exactly 77 (Figure 5). It was therefore a further attempt to reduce the generated vectors. Reduced it to 70. The algorithm in this case also advised and has generated a path with 59 positions (Figure 6). Discretization step is unchanged.

In another example, the increased number of randomly generated positions, in order to check the influence of this number, for example, the shape of the path. For 1000 generated positions the algorithm needed 840 vectors to compute collision-free path of the discretization step 5 [m]. The results present figure 7.

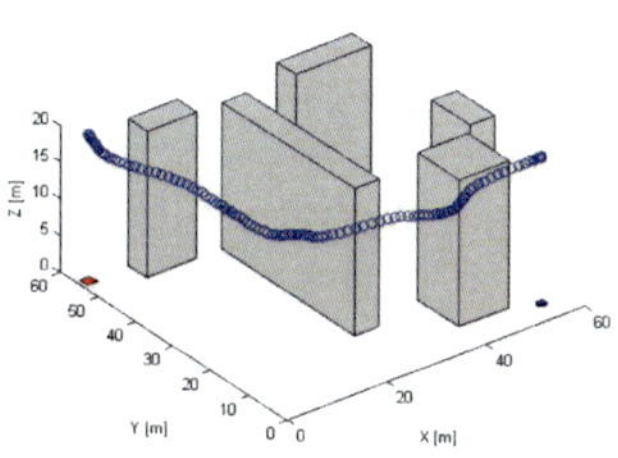

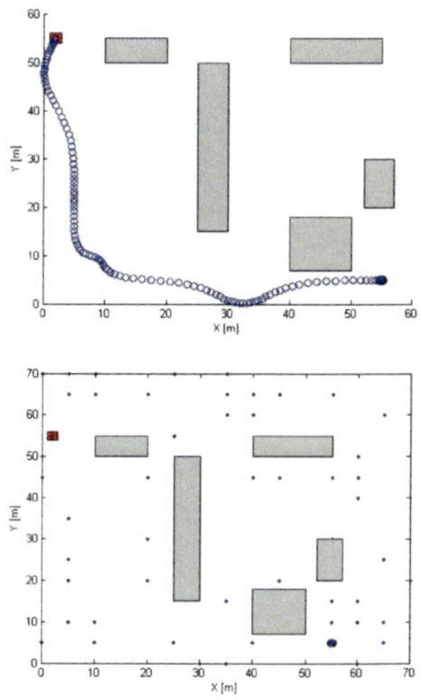

Fig. 6 The simulation: number of position 70, discretization step 5 [m]

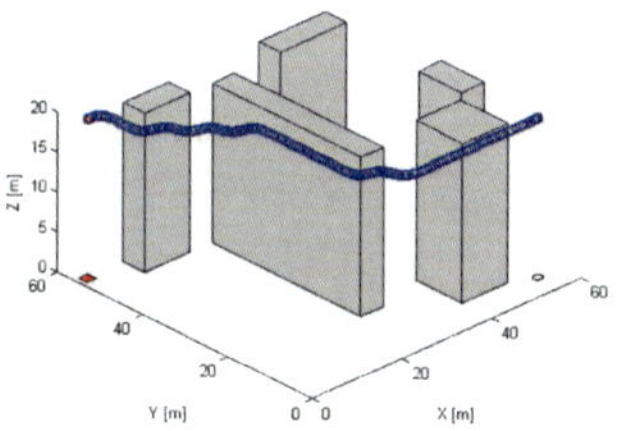

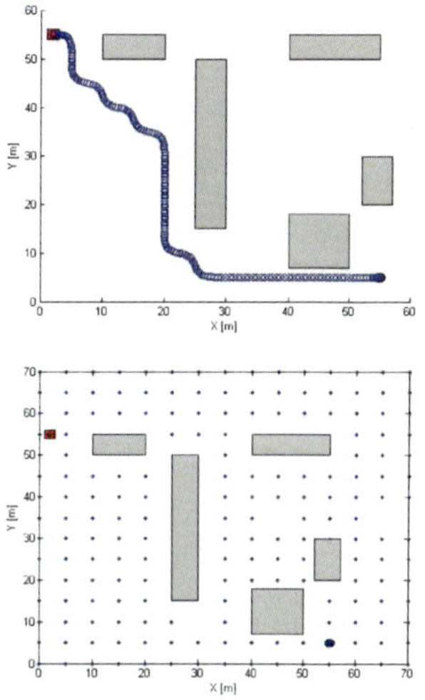

Fig. 7 The simulation: number of position 1000, discretization step 5 [m]

In the following simulations (figure 8, figure 9), the discretization step was reduced to 0.5 [m] and the calculations performed respectively for 100 and 1000 random positions.

The algorithm needed respectively 79 and 815 positions in order to generate the path.

After obtaining satisfactory results of simulation experiments performed for the area shown in the figure 10. The aim is to investigate the effect of density of obstacles in the flight on the effectiveness of planning collision-free path.

As a result of increased obstacles to first-generation collision-free path attempts have failed. Only for 10 [m] discretization step gives a goal. The figure 11 shows

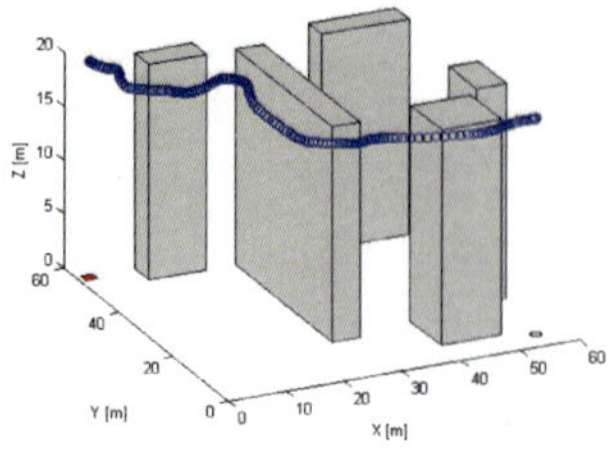
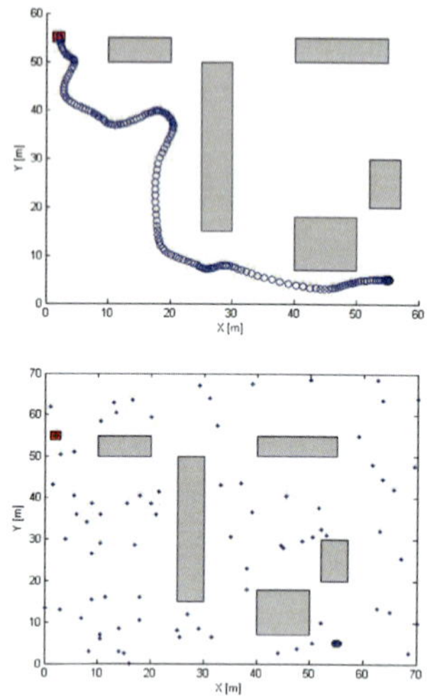

Fig. 8 The simulation: number of position 100, discretization step 0,5 [m]

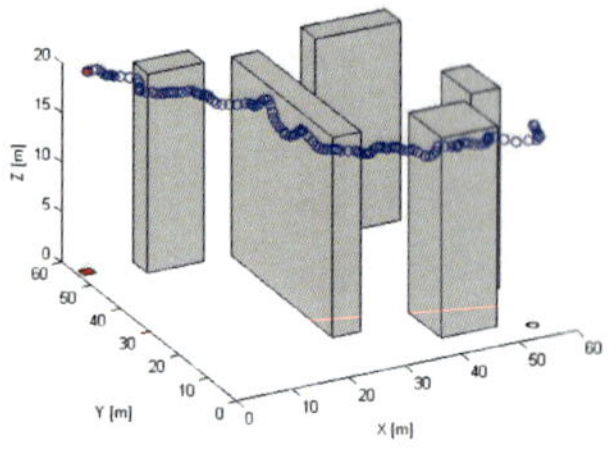
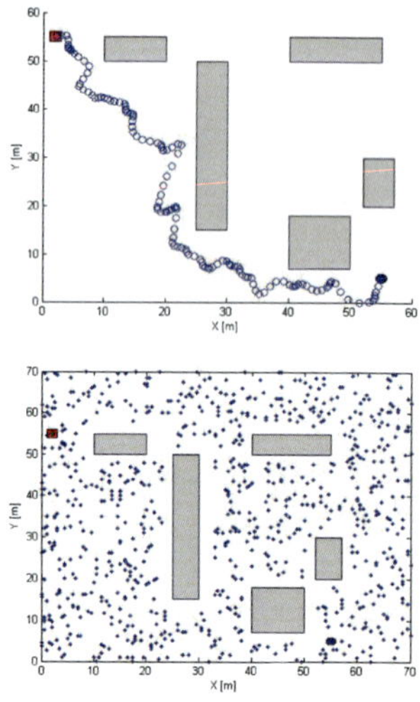

Fig. 9 The simulation: number of position 1000, discretization step 0,5 [m]

the results for 100 sample in which the algorithm calculated the collision-free path generated for 200 positions. The algorithm needed 165 positions.

Just as in the first example, assume that increasing the number of generated positions should improve the algorithm. The results for 1000 random positions illustrated Fig. 12. To calculate the flight path algorithm needed 840 positions.

Despite obtaining collision-free path, shown in the example above, the generation time and number of attempts, after which the obtained results are not fully acceptable for the work control system on-line. In order to improve performance assumes a modification of the proposed algorithm. Precisely in the area of the obstacles created the sub-areas in which additional randomly generated position. This is the density

Fig. 10 The second area

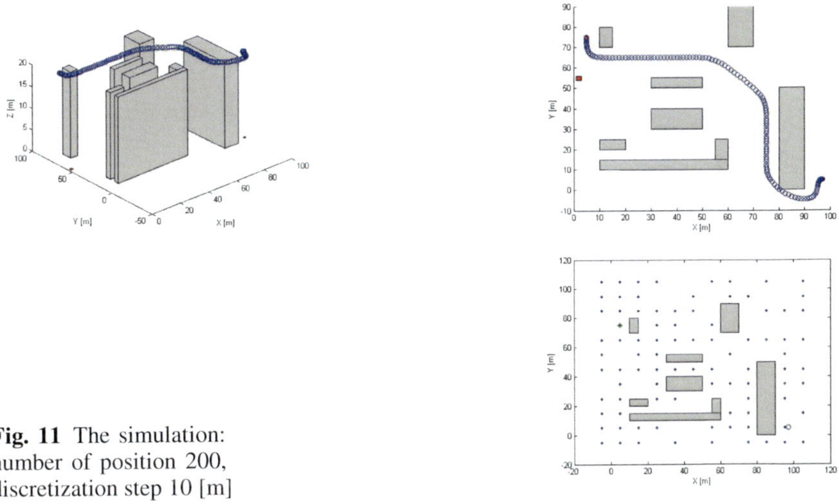

Fig. 11 The simulation: number of position 200, discretization step 10 [m]

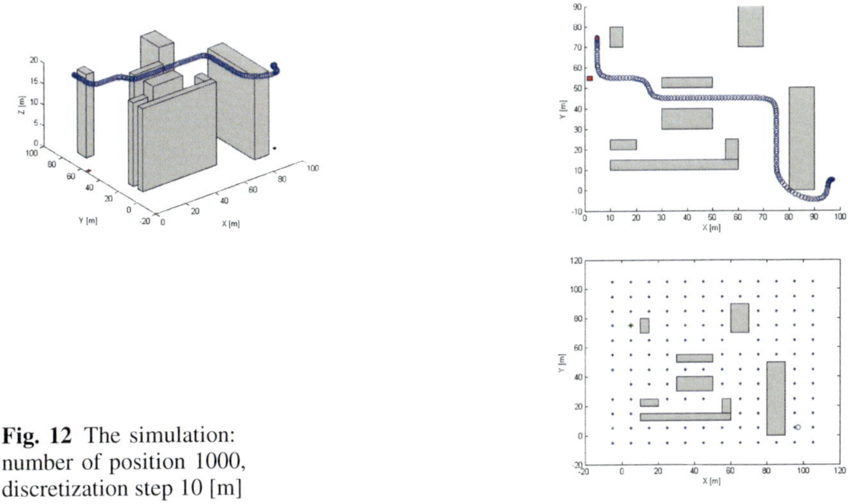

Fig. 12 The simulation: number of position 1000, discretization step 10 [m]

of free space, then already in the standard way to search for collision-free path. Below is an example of executing the path generation in the manner described above. The number of randomized positions in the whole space was set at 200 and the same reference number of positions was generated in each sub-area associated with each obstacle. Whole space discretization step and all sub is 0.5 [m]. The results are shown in the figure 13. The algorithm needed 1250 positions to find a collision-free of the flight path of UAV object.

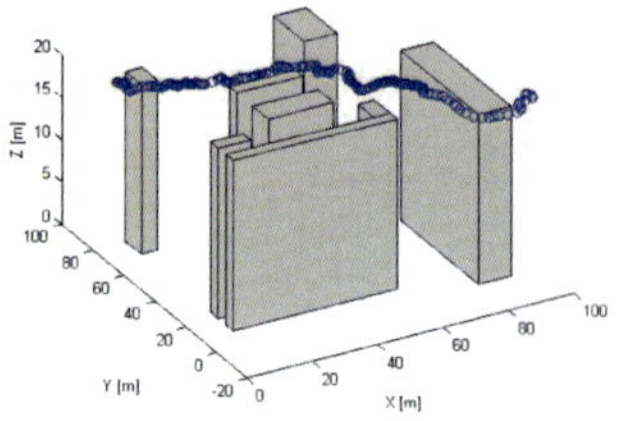

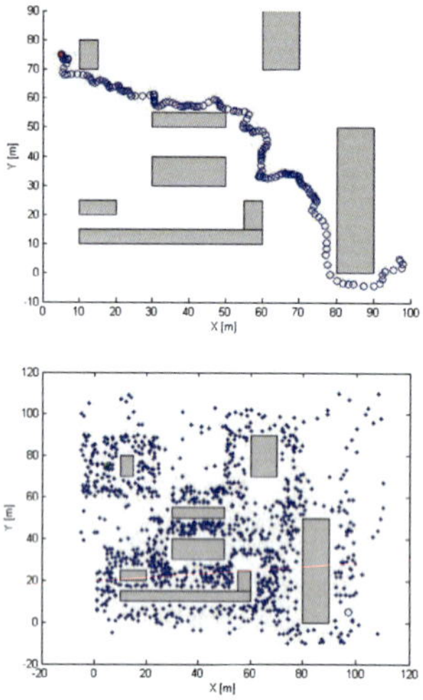

Fig. 13 The simulation using sub-area

5 Summary

The above method is a proposal to solve the problem of planning collision-free flight path for the object type of UAV. In the simulation experiments carried out given that the control system has full information about the area where the helicopter is moving and covering the area obstacles. It is also important to assume that the flying object itself is equipped with fully autonomous autopilot and very good quality digital camera. But the problem is the issue of the calculation. In the above examples assume that the calculations are executed off-line or control system each time waiting for the results of the algorithm. However the main result of the analysis is that it is difficult to find any correlation between the number of preset positions, and the number of selected locations of the collection. The examples are not given the time of calculation, since it also depends on the parameters of the equipment on which the calculations were performed. One can only mention that in the case of calculations for 100 preset setup time was 1 [s], for 1000 of about 2 [s], and for 2000 configurations of about 3 [s]. In addition, simulation experiments carried out do not take into account the criterion for choosing the next location. A very important conclusion after observation of the results is the shape of the flight path. We see that a small number of randomized positions obtained path, which is modeled curves is also spline smooth curve. In some examples, received flight paths are ideal, ie no need to continuously change the direction of flight, the object passes along the obstacles without unnecessary movements, which can be seen in other simulations. Serious problem is the so-called passage through narrow corridors. Well this was illustrated in Example 2, where he concentrated in the area of flight obstacles. Results are presented here only the results of simulation experiments. The correctness of the exact same method can be confirmed through verification of the real object. The main issues that arise are likely to be trouble with the speed of microprocessor computing unit flying object, and image processing to obtain the position of any obstacles. You should also pay attention to the problems of processing the location signals to real signals relevant regulations of UAV systems. This issue is a completely separate issue, and this work has not been touched. But as regards the implementation of the method of B-PPT is characterized by the simplicity of the algorithm and has a very simplified mathematical tools, which uses only the necessary information. The problem is the lack of repeatability of results. This method can be used only if we are to achieve a final position or configuration, since the shape of the path between these points we do not have flow. This is due to base B-PPT method, the use of probability calculus.

References

1. Aarno, D., Kragic, D., Christensen, H.: Artificial potential biased probabilistic roadmap method. In: Proc. IEEE International Conference on Robotics and Automation, New Orleans, pp. 461–466 (2004)
2. Babiarz, A.: Nowe kryterium przeszukiwania w metodzie bazujacej na algorytmie PRM. In: XV Krajowa Konferencja Automatyki, Warszawa, pp. 289–294 (2005)
3. Babiarz, A.: Planowanie trajektorii manipulatorów z zastosowaniem krzywych B - sklejanych. Rozprawa doktorska, Gliwice (2005)
4. Beard, R.W., McLain, T.W., Goodrich, M.A., Anderson, E.P.: Coordinated target assignment and intercept for unmanned air vehicles. IEEE Transactions on Robotics and Automation 18(6), 911–922 (2002)
5. Burns, B., Brock, O.: Information Theoretic Construction of Probabilistic Roadmaps. In: Proc. of the International Conference on Robotics and Systems, Las Vegas, Nevada (2003)
6. Hsu, D., Jiang, T., Reif, J., Sun, Z.: The Bridge Test for Sampling Narrow Passages with Probabilistic Roadmap Planners. In: IEEE International Conference on Robotics and Automation, Taipei, Taiwan, pp. 4420–4426 (2003)
7. Hsu, D., Latombe, J.C., Kurniawati, H.: On the Probabilistic Foundations of Probabilistic Roadmap Planning. In: 12th Int. Symp. on Robotics Research, San Francisco (2005)
8. Kavraki, L.E., Latombe, J.C.: Probabilistic Roadmaps for Robot Path Planning. In: Gupta, K., del Pobil, A. (eds.) Practical Motion Planning in Robotics: Current Approaches and Future Directions, pp. 33–53. John Wiley (1998)
9. Kiciak, P.: Podstawy modelowania krzywych i powierzchni. WNT, Warszawa (2000)
10. Kuffner, J.J., LaValle, S.M.: RRT-Connect: An efficient approach to single-query path planning. In: Proc. IEEE Int'l Conf. on Robotics and Automation, San Francisco, CA, pp. 995–1001 (2000)

11. Kuffner, J.J., Nishiwaki, K., Kagami, S., Inaba, M., Inoue, H.: Motion planning for humanoid robots under obstacle and dynamic balance constraints. In: Proc. IEEE Int'l Conf. on Robotics and Automation, Seoul, Korea, pp. 692–698 (2001)
12. LaValle, S.M.: Rapidly-exploring random trees: A new tool for path planning. Computer Science Dept. (1998)
13. LaValle, S.M., Kuffner, J.J.: Randomized Kinodynamic Planning. Int'l Journal or Robotics Research 5(8) (2001)
14. Nieuwenhuisen, D., Kamphuis, A., Mooijekind, M., Overmars, M.H.: Automatic Construction of High Quality Roadmaps for Path Planning. Utrecht University, technical report UU-CS-2004-068 (2004)
15. Pettersson, P.O.: Helicopter Path Planning using Probabilistic Roadmaps, Master's thesis, Dept. of Computer Science at Linkopings Universitet (2003)
16. Pettersson, P.O.: Sampling-based Path Planning for an Autonomous Helicopter. Linkoping Studies in Science and Technology Thesis No. 1229 (2006)
17. Proud, A.W., Pachter, M., D'Azzo, J.J.: Close formation flight control. In: Proceedings of the AIAA Guidance, Navigation and Control Conference, Portland, OR, AIAA Paper No. 99-4207 (1999)
18. Roskam, J.: Airplane Flight Dynamics and Automatic Flight Controls, Design. Analysis and Research Corporation, Lawrence, KS (2001)

Planning as Artificial Intelligence Problem – Short Introduction and Overview

Adam Gałuszka, Marcin Pacholczyk, Damian Bereska, and Krzysztof Skrzypczyk

Abstract. Planning belongs to fundamental AI domains. Examples of planning applications are manufacturing, production planning, logistics and agentics. Over the decades planning techniques were improved and now they are able to capable real environment problems in the presence of uncertain and incomplete information. This article introduces the notion of so called classical planning, indicating connected with this computational complexity problems and possible ways of treating uncertainty.

Keywords: artificial intelligence, planning, uncertain and incomplete information.

1 Introduction

Artificial Intelligence is a study of design of intelligent agents. An intelligent agent is a system that acts intelligently on its environment. There are various problems which are being investigated by Artificial Intelligence, like knowledge, reasoning, learning and planning [24]. Planning is a task of coming up with a sequence of actions that will achieve a goal. Finding an optimal plan is generally a hard computational problem and needs a lot of resources. The situation becomes even more complicated when a planner does not have a complete set of information about an environment for which the plan should be created. This is called uncertainty and is essential for exact description of a real environment. There exist large number of different approaches and heuristics that try to deal with planning with uncertainty depending on its kind.

Adam Gałuszka · Marcin Pacholczyk · Damian Bereska · Krzysztof Skrzypczyk
Silesian University of Technology, Institute of Automatic Control,
Akademicka 16, 44-101 Gliwice, Poland
e-mail: {Adam.Galuszka,Marcin.Pacholczyk}@polsl.pl,
{Damian.Bereska,Krzysztof.Skrzypczyk}@polsl.pl

A. Nawrat et al. (Eds.): Advanced Technologies for Intelligent Systems, SCI 440, pp. 95–103.
springerlink.com

Planning should be distinguished from *scheduling* - well-known and frequently used technique of improving the cost of a plan. Planning is understood as causal relations between actions, while scheduling is concerned with metric constraints on actions [1]. When all states of a planning problem (including an initial and a goal) are described by a given set of conditions (also called predicates), then the problem is called STRIPS planning problem [22]. There are many applications of the planning problems in industrial processes, production planning, logistics and robotics [1]. The STRIPS system has been successfully applied in planning modules of Deep Space One Spacecraft [29] and for elevators control in Rockefeler Center in New York [16]. Planning and its computational aspect takes also much attention in the AI community [2, 3, 5, 7, 11, 14, 23, 24].

2 STRIPS Planner

STRIPS (STanford Research Institute Problem Solver) is the planner developed in 1971 by Fikes and Nilsson from Stanford Research Institute, Menlo Park, California. It was called a 'problem solver' but planner is a more accurate name in the modern Artificial Intelligence terminology.

This planner was originally implemented in Lisp and was used in the 'Shakey the robot' project [27]. Its aim was to develop an autonomous robot which could act in the well defined world, called 'World of Blocks' in its simplified form. Shakey was supposed to operate in his closed world, switch lights on and off, push and climb boxes and open and close doors. It should be mentioned that 'STRIPS' is a name used to describe both, the original Strips planner and the Strips representation language. Even if the Strips planner is not used anymore, the Strips language is widely used to describe worlds and operators in large number of traditional, but more advanced, planners.

As stated in [21] 'the task of problem solver is to find some of the operators that transforms a given initial world model into one that satisfies some particular goal condition' and the task of Strips system in the 'Shakey the robot' project was to 'find a sequence of operators that would produce a world model in which the goal could be shown to be true'.

Both operators and world model descriptions were stated as first-order predicate calculus well formed formulas, or wffs. A well formed formula is 'first order predicate calculus sentence that has all its variables properly introduced [...] before it is used' [9]. The instance of propositional Strips planning is specified by four sets (P, O, I, G), where:

- P - is a finite set of ground atomic formulas called the conditions or properties.
- O - is the finite set of actions called operators. It consists of list of:
  - Precodnitions - define what must be established before the operator can be applied. It is a conjunction of positive (true) and negative (false) conditions of the operator.

  - Postconditions - define the effects of applying the operator. It is a satisfiable conjunction of positive and negative conditions (called postconditions) of the operator. They show which conditions are made true and which are made false after applying an operator. For convenience, the postcondition list is usually shown as two lists: an add list and delete list, representing conditions to be added and deleted after applying an operator, respectively.

- I - describes an initial state. It is a set of conditions that are initially true.
- G - specifies which conditions are true and false in order for a state to be considered as a goal. The goal state is a satisfiable conjunction of positive and negative goals [6, 21].

An action (operator) can be applied to any state, but has an effect only if its preconditions are satisfied. Otherwise, the action has no effect. If the preconditions of an operator are satisfied, its positive literals representing effects (postconditions) are added and its negative literals representing effects (postconditions) are deleted. If the positive effect is a part of the world's description before applying an operator it is not added twice. Similarly, if the negative effect is not present in the world description before applying an operator, the part of the effect is ignored. This is so called Strips assumption: every literal not mention in the effect remains unchanged.

The solution of the planning problem is an action sequence that, when executed in the initial state, results in state that satisfies the goal [26].

As was stated above, 'problem solving means finding a sequence of operators in the space of world models that will transform an initial model into another model in which a given formula can be proven to be true' [9]. Strips used two main techniques for problem solving:

Means end analysis - used for goal search
Resolution theorem prover - used for answering questions about a given world model.

The world model in Strips is perfectly known in each step of planning. This knowledge, together with knowledge of both, pre- and post-conditions of operators allows one to apply both forward and backward search methods. Forward search method searches from initial to goal model while backward search method searches from goal to initial model. The means-end analysis combines many of the advantages of both search methods. It can be informally described as: 'If the difference between what you have and what you want is of this kind, then try to reduce that difference by this method' [9]. The search through situation spaces can be summarized as follows, from [9]:

1. Extract the 'differences' between the current world model and the goal
2. The theorem prover identifies the 'relevant' operators for reducing these differences.
3. Solve the subproblem of producing a world model where a relevant operator can be applied.
4. Repeat until all goals are achieved or no operator can be applied.

Theorem prover in Strips is used for:

- Proving that goal is true in a given set of clauses
- Determining the operator's applicability in a given world model
- Computing the 'differences' between current state and a goal
- Selecting 'relevant' operators to reduce the difference

The prove that a goal G(p), (where p is a set of parameters) is true in a given world model M is done by proving the inconsistency of the set $\{M \cup \neg G(p)\}$ and more specifically, by finding such instance p', where $p' \subseteq p$, for which the set $\{M \cup \neg G(p)\}$ is logically inconsistent. The operator's applicability in a given world model is also done by the presented procedure, however this time the G(p) represents the operator's preconditions. The 'difference' $P = M - G_i$ between current situation M and the current subgoal G_i is attached to the current node if G can not be proved in M. In this situation, the 'relevant operators' are those that will allow continuation of P. They can be found in two steps:

- Firstly, the candidate operators are found. These are operators that contain any of the predicates of P in their positive postcondition's list (add list).

The operator is called 'relevant' for the given difference if the clauses that it adds can be used to continue the proof P.

3 Partial-Order Planning

The original Strips planner is a linear planner and creates totally ordered plans. It means that it searches in a spaces of world models and represents plans as totally ordered sequence of actions, one that is directly connected to the start and goal. Strips could not decompose the problem and work on each sub-problem separately, in contrary, it had to make decisions on how to sequence the actions from all the sub-problems at once [26].

This was rather inconvenient method. It would be preferred to divide a problem to several subproblems, create a subplan for each of them separately and then combine them into one plan.

Here we introduce the problem of partial-order planning together with its most important definitions. They are based on a Strips language which can be adjusted to handle partial-order planning.

Partial order planner is a planning algorithm which can place two actions in a plan without specifying which comes first. Planner performs a search in the space of partial-order plans. It begins with empty plan and consider ways of refining it. It is done until the complete plan which can solve the problem is found. The actions in this search are actions in plan and not in the world. The Start and Finish are called dummy actions. They are called actions because the complete plan will be made only from actions.

Each partial-order plan is composed from the following sets, taken from [26]:

- *Actions*. They make up the steps of plan and are taken directly from set of actions in the planning problem.
- *Ordering constraints*. They introduce an ordering in actions by describing proper partial order. An ordering constraint is of the form $A \prec B$ what means that 'A is before action B'. It does not have to be immediately before. A *contradiction* is a state when ordering constraints make a cycle, such as: $A \prec B$ and $A \prec B$. An ordering constraint can not be added to the plan when it creates a cycle.
- *Causal links*. They indicate the dependencies among steps in a plan. A causal link is a triple $< A, p, B >$ where p is a proposition symbol, B is a step name that has p as a prerequisite and A is a step name that have p in its add list. A causal link $A \rightarrow^p B$ is read 'A achieves p for B' and requires that step A precede B and that no step between A and B either adds or deletes p [20].
- *Open preconditions*. These are the preconditions that are not achieved by any actions in a plan. The set of open preconditions should be decreased to an empty set by planner, without introducing a contradiction.

A consistent plan is a plan in which there is no cycles in ordering constraints and no conflicts with the causal links [26].

A *solution* is a consistent plan with no open preconditions.

As an example consider a problem of putting on a pair of shoes. The problem is described by:

I: ()

G: $RightShoeOn \bigwedge LeftShoeOn$

O:

RightShoe Precondition: *RightSockOn* Precondition: Add: *RightShoeOn*	***RightShoe*** Precondition: Precondition: Add: *RightSockOn*
LeftShoe Precondition: *LeftSockOn* Precondition: Add: *LeftShoeOn*	***LeftShoe*** Precondition: Precondition: Add: *LeftSockOn*

The partial-order plan has the following components:

Actions: { *RightSock, RightShoe, LeftSock, LeftShoe, Start, Finish* }

Orderings: { $RightSock \prec RightShoe, LeftSock \prec LeftShoe$ }

Open Preconditions: { }

Linking:

$$\{RightSock \xrightarrow{RightSockOn} RightShoe,\ LeftSock \xrightarrow{LeftSockOn} LeftShoe$$

$$RightShoe \xrightarrow{RightShoeOn} Finish,\ LeftShoe \xrightarrow{LeftShoeOn} Finish\}$$

4 Complexity of Planning

Planning is a very complex problem. Most of the interesting classical planning problems are found to be NP-Complete or PSpace complete. It is important to know basic definitions in the field of complexity analysis before continuing the work, therefore (teken from [23, 31]:

Complexity class - set of problems which needs similar computational complexity.

Class P, or *polynomial* class of problems are all problems that can be solved in polynomial time. More formally, these are the problems that can be solved by a Deterministic Turing Machine in polynomial time.

Polynomial time is a computational time of a problem where the run time, m(n), is no greater than a polynomial function of the problem size, n.

Class NP, or *nondeterministic polynomial*, is a set of problems for which there is some algorithm that can guess a solution and then verify whether the guess is correct in polynomial time [26]. More formally, the complexity class NP is the set of decision problems that can be solved by a *non*-deterministic Turing machine in polynomial time [23, 31].

NP-complete problems, a subclass of NP class refers to the 'most extreme' problems of NP. It has been proven that either all NP-complete problems are in P or none of them is. This class contains many important problems.

Class PSpace refers to problems that require a polynomial amount of space even on a nondeterministic machine. It is believed that Pspace hard problems are worse than NP-complete [26].

In general, to determine if a given planning instance of Strips has any solutions is PSpace-complete. The complexity can be however decreased to polynomial time or even NP-completeness by applying different conditions, like decreasing number of pre- and post-conditions [6, 7, 13]. These restrictions, however, limit the applicability of a given plan. Non-classical planners are those where the planner does not have a complete knowledge about the world. In this situation, the planning problem can become even harder computationally. It was shown that the planning problem in presence of incompleteness belongs to higher complexity class as NP-complete, namely NP^{NP} -complete, or Σ_2-complete [2].

It is worth to mention that Bylander in [8] proposed new heuristic for propositional Strips planning which should decrease the complexity of planning to Polynomial class. This method is based on transforming planning instances to linear programming instances. Linear programming methods can be used to solve planning problems after proper mapping of planning instances into set of linear programming variables.

5 Dealing with Uncertainty

The world we live in is full of uncertainties: for example the tomorrow's weather or the world's peace. This is part of people's life and while we grow up we learn, to certain extent, to perceive the uncertain phenomena. We also feel how to use it. As example such sentence can be cited: "the weather is warm today", what can have different meanings whether we are in Africa or in Alaska or is it Summer or Winter in Poland. Such uncertainty can not be measured, but felt and it is called *Cognitive Uncertainty* [12].

Uncertainty arising from physical phenomena, like vibration of an engine or a string can be, however, measured and described in details. This uncertainty arise from random behaviour of physical systems [12] and there are sufficient mathematical tools, developed by centuries, to describe them. Such type of uncertainty is called *Probabilistic*.

The presented 'models' of uncertainties can be treated as two main groups, however there are many different kinds of uncertainties and it only depends on the user and purpose for which it will be used [4]

One of the first planners to handle large problems with many sources of uncertainty is Weaver, created by Jim Blythe [5]. Weaver has developed techniques that allow planner to find plans in situation where there can be many sources of uncertainty and reasons efficiently about these plans. It can handle uncertainty from three types of cause:

- more than one possible initial state
- nondeterministic outcomes of actions
- non-deterministic exogenous events

The third cause of uncertainty is very important because external events can very often play an important role in real domains. Theoretically, other planners can also model them. It could be done by modelling effects of exogenous events by effects of actions in the domain. This is, however, very inefficient way, since their effects are duplicated in every operator in the domain. In such a way each operator might have to model many events.

Weaver is build on Prodigy planner, a domain-independent planner created as test-bench for learning and problem solving which provides a rich language for specifying search control knowledge [5, 28]. As an output, Weaver, similarly as Buridan, gives a plan exceeding desired minimum probability of success. In order to avoid combinatorial explosion of explicit consideration of each of all possible alternative outcomes, Weaver uses similar approach to Buridan, namely build a plan using two alternating steps: plan creation and plan evaluation. In the first part, an outcome for a step that is desired in order to achieve a goal is chosen. The step does not, explicitly, consider the other outcomes, however it makes sure the outcome is reasonably likely. In the next step, plan evaluation, all sources of uncertainty can be potentially considered, however Weaver explicitly represents only those that are shown to influence the plan's chance of success. This is very important feature of this planner since taking the relevant sources of uncertainty can lead to significant computational savings. Then, the plan creation is re-entered with one or more sources of uncertainty that are known to be important are being explicitly considered [5].

6 Conclusions

Planning problem is a complicated computational task, even in case were knowledge about environment is complete. Introducing uncertain and incomplete information to model of the problem, computational complexity usually drastically increases. It is main reason, one should carefully choose the right technique to be able to find problem solution in reasonably time. But if proposed methodology succeed, then artificial intelligence system is powerful tool: basing on sensor information that models current state of planning problem, and given desired goal situation, system can determine whether the solution exists and, in this case, produce plan that allows achieve the goal.

References

1. Backstrom, C.: Computational Aspects of Reordering Plans. Journal of Artificial Intelligence Research 9, 99–137 (1998)
2. Baral, C., Kreinovich, V., Trejo, R.: Computational complexity of planning and approximate planning in presence of incompleteness. In: Proc. 16th International Joint Conference on Artificial Intelligence, IJCAI 1999 (1999)
3. Barret, A., Weld, D.S.: Partial-Order Planning: Evaluating Possible Efficiency Gains. Artificial Intelligence 67, 71–112 (1994)
4. Blythe, J.: An Overview of Planning Under Uncertainty. Pre-print from AI Magazine 20(2), 37–54 (1999)
5. Blythe, J.: Planning Under Uncertainty in Dynamic Domains. Ph.D. Dissertation. Carnegie Mellon University Computer Science Department
6. Bylander, T.: The Computational Complexity of Propositional STRIPS Planning. Artificial Intelligence 69, 165–204 (1994)
7. Bylander, T.: A probabilistic analysis of propositional STRIPS planning. Artificial Intelligence 81, 241–271 (1996)
8. Bylander, T.: A linear programming heuristic for optimal planning. In: Proceedings of the 14th National Conference on Artificial Intelligence, pp. 694–699 (1997)
9. Cocosco, C.A.: A review of STRIPS: A new approach to the application of theorem proving to problem solving by R.E. Fikes, N.J. Nillson, 1971. For 304-526B Artificial Intelligence 4 (1998)
10. Cormen, T.H., Leiserson, C.E., Rivest, R.L.: Wprowadzenie do algorytmów. WNT, Warszawa (1997)
11. Gerevini, A., Schubert, L.: Accelerating Partial Orders Planners: Some Techniques for Effective Search Control and Pruning. Journal of Artificial Intelligence Research 5, 95–137 (1996)
12. Gupta, M.M.: Intelligence, uncertainty and information. In: Machine Intelligence and Pattern Recognition, Analysis and Management of Uncertainty: Theory and Applications, vol.13. North-Holland (1992)

13. Gupta, N., Nau, D.S.: On the complexity of Blocks - World planning. Artificial Intelligence 56(2-3), 223–254 (1992)
14. Haddawy, P., Suwandi, M.: Decision-Theoretic Refinement Planning using Inheritance Abstraction. In: Hammond, K. (ed.) Proc. Second International Conference on Artificial Intelligence Planning Systems. University of Chicago, AAAI Press, Illinois
15. Klir, G.J., Folger, T.A.: Fuzzy Sets, Uncertainty, and Information. Prentice-Hall International, Inc., USA (1988)
16. Koehler, J., Hoffmann, J.: On Reasonable and Forced Goal Orderings and their Use in an Agenda-Driven Planning Algorithm. Journal of Artificial Intelligence Research 12, 339–386 (2000)
17. Kushmerick, N., Hanks, S., Weld, D.: An algorithm for probabilistic least-commitment planning. In: Proc. Twelfth National Conference on Artificial Intelligence, pp. 1073–1078. AAAI Press (1994)
18. Leckie, C., Zukerman, I.: Inductive Learning of Search Control Rules for Planning. Artificial Intelligence 101, 63–98 (1998)
19. Long, D., Fox, M.: Efficient Implementation of the Plan Graph in STAN. Journal of Artificial Intelligence Research 10, 87–115 (1999)
20. McAllester, D.: Nonlinear Strips Planning. Lecture notes for Massachusetts Institute of Technology's course: 6.824, Artificial Intelligence (1992)
21. Nillson, N.J., Fikes, R.E.: STRIPS: A new approach to the application of theorem proving to problem solving. Technical Note 43, SRI Project 8259, Artificial Intelligence Group, Stanford Research Institute (October 1970)
22. Nillson, N.J.: Principles of Artificial Intelligence. Toga Publishing Company, Palo Alto (1980)
23. Papadimitriou, C.: Computational Complexity. Addison Wesley (1994)
24. Poole, D., Mackworth, A., Goebel, R.: Computational Intelligence: A Logical Approach. Oxford University Press (1998)
25. Rintanen, J.: Constructing Conditional Plans by a Theorem-Prover. Journal of Artificial Intelligence Research 10, 323–352 (1999)
26. Russell, S.J., Norvig, P.: Artificial Intelligence: A Modern Approach, 2nd edn. Prentice Hall Series in Artificial Intelligence (2003)
27. Stanford Research Institute: Shakey the robot, `http://www.ai.sri.com/shakey/`
28. Veloso, M., Carbonell, J., Perez, A., Borrajo, D., Fink, E., Blythe, J.: Integrating planning and learning: The prodigy architecture. Journal of Experimental and Theoretical AI 7, 81–120 (1995)
29. Weld, D.S.: Recent Advantages in AI Planning. Technical Report UW-CSE-98-10-01; also AI Magazine (1999)
30. Weld, D.S., Anderson, C.R., Smith, D.E.: Extending Graphplan to Handle Uncertainty & Sensing Actions. In: Proc. 15th National Conf. on AI, pp. 897–904 (1998)
31. Wikipedida, a free encyclopedia, `http://en.wikipedia.org` (Cited July 18, 2008)

Part III
Sensoric and Vision Based Information for UAVs

One of the most important and challenging aspects of both control and application of UAV is sensoric and vision based information. Acquisition of multimodal data from the sensors mounted to the UAV allows the mechanical elements of the object to follow the computed trajectory and, as a result, collision-free navigation. Before the data are entered to the navigation it requires selection, dimensionality reduction and filtering. One of the important topics is accurate calibration of the sensors, especially the inertial measurement units (IMUs). In the chapter, a new calibration method of IMU is presented. It allows calibration of multiple IMU sensors at the same time with the same resulting quality, which is a novelty of a considerable importance. Another important topic is fusion of the output signals from the altimeters, IMU and GPS systems. The result of the fusion can be successfully applied to solving navigation problems. Common obstacles avoidance is implemented on the basis of various range finders. Here, the results of studies of the Hokuyo UTM-30LX Laser Range Finder are presented. The study included obstacles avoidance in a dynamically changing environments. It is experimentally proved that the tested range finder is suitable for use in autonomous mobile robots and can be used for distance scanning, targeting different materials, colors and brightness with similar effects. The relative errors of measurements are below 1.

Nowadays, UAVs are mostly used for observation from the air. Therefore, vision and image processing algorithms are inseparable from the UAVs. Depending on the length of magnetic waves perceived different information can be acquired. There are visible light cameras, low-light cameras, infrared cameras, noctovision cameras and multi and hyperspectral cameras capable of mounting on board of the aerial vehicles. In the chapter a study and comparison of all types of cameras are presented. They were tested during both day and night in order to measure the possibility of application for night time surveillance missions. After selection of the most promising for future processing types of cameras, a multispectral image acquisition 2-DOF platform (gimbal) was designed and constructed. Finally, using image processing algorithms, the video stream from both IR and visual camera were used in order to detect and track the targets. Acquired results of the mentioned algorithms are presented and discussed in detail.

The Prototype of Gyro-Stabilized UAV Gimbal for Day-Night Surveillance

Karol Jędrasiak, Damian Bereska, and Aleksander Nawrat

Abstract. This paper presents a designed and created prototype of stabilized UAV gimbal for day-night surveillance. There is a need for UAV gimbals capable of stable target detection and tracking, regardless the time of the day. The prototype gimbal is capable of pan and tilt rotations in range of full 360 degrees with 10 bit turning resolution. Full turn takes approximately 1.2s. The gimbal itself and possible future external devices are able to communicate via extended VISCA protocol and RS-232C standard. Performed tests and evaluations have shown promising results regardless the time of the day. The presented gimbal shows a great potential for target detection and tracking using UAVs and developing more sophisticated algorithms for stabilization.

Keywords: surveillance, gimbal, thermal imaging, UAV stabilization.

1 Introduction

Unmanned aerial vehicles (UAVs) are often used for infrastructure and strategic building surveillance. Recently they are not only used for military applications but also for geodesic measurements and mass events surveillance. Due to the requirement of minimal size, weight and energy usage, micro UAVs are equipped only with the most necessary, often simplified, devices. In this paper we present a designed and created working prototype of Gyro-Stabilized Gimbal for day-night surveillance applications.

Karol Jędrasiak · Damian Bereska · Aleksander Nawrat
Silesian University of Technology, Institute of Automatic Control,
Akademicka 16, 44-101 Gliwice, Poland
e-mail: {Karol.Jedrasiak,Damian.Bereska}@polsl.pl,
Aleksander.Nawrat@polsl.pl

A. Nawrat et al. (Eds.): Advanced Technologies for Intelligent Systems, SCI 440, pp. 107–115.
springerlink.com

Detection and tracking algorithms are sensitive to vibrations caused by propulsion system of the flying objects as well as the influence of the environment e.g. wind. It is important to reduce the impact of those vibrations on the results of object tracking and remote sensing UAV's capabilities. Video stream stabilization regarding the video camera mounting can be achieved by stabilization at the level of the whole aerial vehicle [1], software stabilization level [2] or at gimbal level. Software image stabilization is computationally complex, therefore it is not wise to perform it unnecessary by UAV's CPU. PID or fuzzy controllers are often used for stabilization with the usage of servos. Recently fuzzy PID composite control was introduced [3].

Video cameras and image processing techniques are a basis of visual surveillance. However vision information is used not only for target tracking but also for navigation e.g. estimating the attitude of the UAV or horizon detecting [4]. Typically, there is mounted a single video camera for day applications [5]. The camera can be hard mounted or capable of pan, tilt rotations to follow the object of interest. Acquired aerial images are later often georegistered and used for creation of a detailed map of the area or image mosaic [6].

Designed gimbal, in order to perform tasks regardless of the day/night conditions, is equipped with two video cameras. First is a visible light video camera used for day applications. Second is a thermal imaging camera used for additional information during day and object detection and tracking during night. Thermovision camera can be used for object detection, localization and tracking as presented in [8]. The prototype pan-tilt gimbal can further improve capabilities of object tracking by rotation in adequate direction. In order to improve the speed of computations image processing algorithms can be implemented in FPGA as in [9].

Visible and thermal video fusion can also be used in order to improve detection and tracking quality. Regions of interest found in thermal images are capable of using, as an additional source of information, for e.q. particle filter tracking as presented in [10]. Fusion of both streams can also be presented to a human operator at the ground station. Such fusion can show additional details in a single video stream and therefore aid the task of monitoring especially during twilight.

2 Gimbal Overview

Continuous object tracking, regardless of the UAV movements, is an important task for the gimbal. Our prototype gimbal (fig. 1) is capable of pan and tilt rotations in full 360 degree range. Two servo mechanisms with ten bit resolution are used for turning. Pan and tilt precision is therefore equal to 0,35 degree. Full 360 degree turn is performed in exactly 1.2 s. It was decided that first prototype is to be restricted to single turn only.

To reduce the impact of vibrations and environmental forces on video cameras streams, a 9-DOF (degree of freedom) IMU (inertial measurement unit) was mounted inside the prototype gimbal. There are 3-DOF of gyroscopes of X, Y and Z axes and respectively 3-DOF of linear accelerometers. Additionally, there are also

3-DOF magnetometers for computing reference data. At the current state of development, data from gimbal's IMU is used for hardware gimbal level stabilization. It is based on the difference between current servos position and the position set. Detected difference is minimized in real time. Gyroscopes measurements were tested and would be used in next implementation to minimize using servos at an angle the gimbal was rotated. In next implementations IMU's data could be used for more sophisticated hardware stabilizations algorithms or by ground station for visualizing current gimbal state.

Fig. 1 The prototype gimbal

Except the autonomous hardware stabilization, additionally, the gimbal can be remote controlled by a PC computer. This feature can be used for tracking detected target using pan-tilt gimbal's capabilities without the need to change the trajectory of the whole UAV.

3 Gimbal Construction

The created and tested prototype of the stabilized optical gimbal is shown in fig. 2. The gimbal is a result of the first stage of design and construction universal stabilized platform for UAV. Aim of the work was to produce a stabilized platform that was capable of installing visible light and IR camera. Priority was to enable rotations around axes x and z.

Size of the empty space was designed for installing a various type of additional sensors and video cameras. It was optimized for easy installing most of the visible light and IR camera models. Gimbal's cylindrical shape is large enough for: two video cameras, propulsion system of the axis x, IMU and a required electronic chips

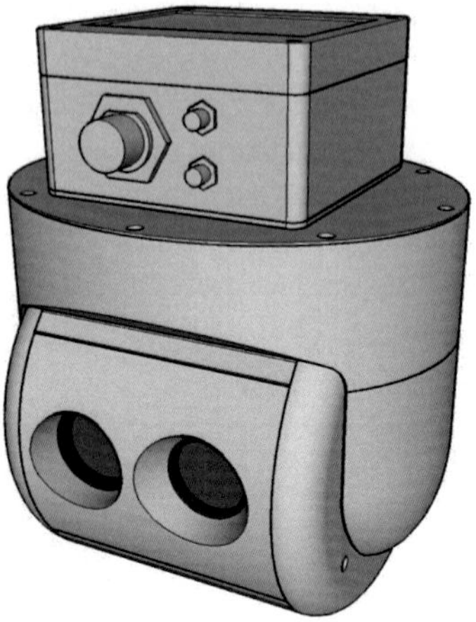

Fig. 2 The gimbal visualization

for control and communication. An additional space for possible future sensors was also designed at the mounting part of the gimbal.

Applied propulsion system enables rotation in the range from 0 to 360 degrees in both axes. It was decided to limit the range to the single full rotation and avoid using rotor connector in order to simplify the communication with the installed sensors and video cameras. Designed range is assumed to be sufficient since target UAV type is a rotorcraft which can easily change vertical orientation. Capability of full rotation around x axis was designed in order to allow hiding cameras in a safe zone of the gimbal. Such an approach significantly reduces the risk of any damage and allows installing video cameras lenses pneumatic cleaning system.

4 Communication with External Devices

The prototype gimbal is communicating with external devices through RS 232 standard and extended VISCA protocol developed by Sony Corporation. Parameters of the RS-232C communication are as follows [11]:

- communication speed: 9600 bps,
- 8 data bits,
- 1 start bit,
- 1 stop bit,
- no parity,
- no flow control.

There can be multiple (up to 7) external devices in VISCA. Each device is identified by its number starting from 1. PC computer is a controller with id equal to 0. Sending

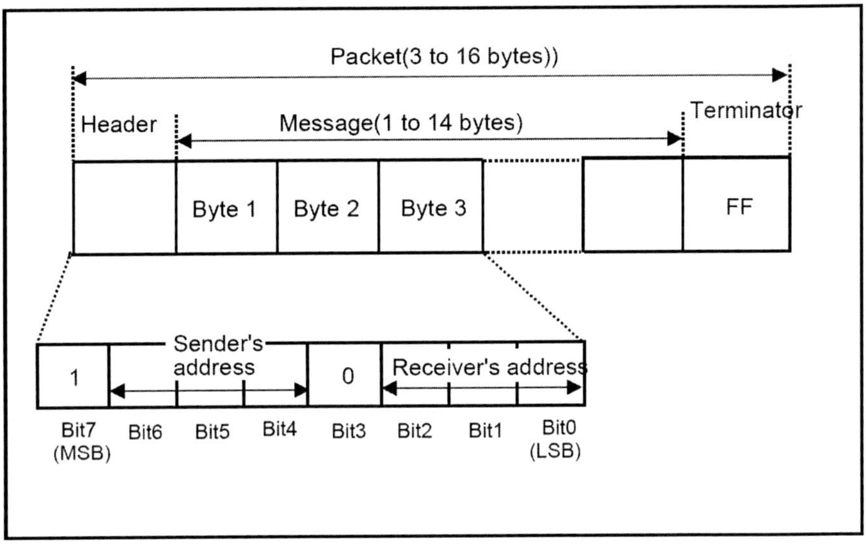

Fig. 3 VISCA packet structure

a command with an id of 8 is interpreted by VISCA as the broadcast command to all devices. Commands are send in data packets (fig. 3) consisting of 1 byte header, up to 14 bytes of command data and an end of packet, byte 0xFF.

Before communication a controller has to be initialized by sending an init command. After initialization, gimbal's servo manager is available. Each of servos is controlled with a command consisting of bytes: 0x81, 0x01, A, B, C, D, S, 0xFF.

S is the servo mechanism speed. It is a value in range between 0x00 and 0x80. A and B, C and D are respectively lower and upper bytes of the number in range from 0 to 1023. A and B are used to control pan servo while C and D values are used for tilt servo mechanism. Lower and upper bytes are computed using the equations (1-2):

$$U_b = value\, div\, 32, \tag{1}$$

$$L_b = value\, mod\, 32, \tag{2}$$

where:

U_b, upper byte value,

L_b, lower byte value.

Video streams from the cameras mounted in the gimbal can be accessed using identical standards as the original ones e.g. PAL or 384x288 pixels resolution.

5 Tests during Day and Night

Experiments were designed in order to test gimbal's surveillance capabilities. Test goal was to detect and track object of interest. Object with temperature that significantly varies (parameter alpha) from the background temperature was assumed as an object of interest. All tests were performed during day and night on a low class PC computer equipped with a 1,8 MHz and 2GB of RAM.

In the scene observed during experiments there are three physical objects. The object of interest is a bottle of hot water. It is placed on top of the table in the room temperature. In the background a wall is visible. During experiments gimbal was moved or rotated and its capabilities to track the object were recorded.

At first, try standard motion detection using background modeling was used to detect object of interest. Image frame was divided into segments and each segment was continuously updated using equation (3):

$$B_{t+1} = \alpha I_t + (1 - \alpha) B_t, \tag{3}$$

where: B_{t+1}, segment background value,

I, segment median intensity,

α, speed of background update.

However, due to thermal camera internal image equalization, the algorithm was instable. It was decided to use image global average as a reference value for thresholding. Each pixel with intensity value different then average background value by beta parameter was assumed as a foreground pixel. Beta parameter was experimentally set as 0.3.

All detected pixels were grouped into blobs and the blob characterized by the largest area was chosen as the successfully detected object. Object tracking was implemented as follows: first, blob is detected in the frame, next, in the second frame, blobs are detected and compared against the list of blobs found in the previous frame. If Euclidean distance of the blobs's center of mass is smaller than DELTA parameter, then it is assumed that the two virtual objects represent the same physical object.

Detected object's distance from the center of the image frame was used as a control signal for servo mechanisms. If the distance along x and y axes is larger than gamma parameter, then servo mechanisms are turning in the direction with the speed proportional to the blob's distance from the image's center.

Acquired results during tests are presented in the following section.

6 Experiments Results

During the first test, the object of the interest was placed in the scene. The test was performed during the day as all the following tests. In laboratory conditions there was no difference between thermal camera video streams from day and night. Therefore, only day results are presented. Successfully detected object is bound with a yellow rectangle. The green line connects the center of mass of the object and the center of the image frame. The line represents distance of the object from the center. In fig. 4A it can be seen that the object is in a distance of about quarter of the image center. Measured distance is used to control servos in order to rotate the gimbal that the detected object is in the middle of the thermal camera image frame. Screenshot images from the turning process are visible in fig. 4B and fig. 4C.

Fig. 4 Images acquired during first experiment

During the second test, gimbal was rotated by a yaw angle and translated to the left. Test's goal was to check whether gimbal was able to detect and track object during simulated UAV movement. As shown in fig. 5 the prototype gimbal fulfilled the test.

Fig. 5 Images acquired during the second experiment

The last, third test's goal was to track object during gimbal rotation by a roll angle. In fig. 6. it can be seen that gimbal successfully tracked object during rotation however, a lack of degree of freedom for roll angle resulted in a rotated image.

During UAV operation usually there is no need for 3-DOF because observed objects are relatively small in comparison to the whole image frame, therefore objects rotations do not affect the tracking process.

Fig. 6 Images acquired from thermal imaging camera during rotation by a roll angle

7 Conclusions

In this paper the prototype of stabilized UAV gimbal for day-night surveillance was presented. Firstly, the problem of detection and tracking during night was introduced. Next, the stabilization problem and possible solutions were mentioned. The designed and created gimbal is capable of pan and tilt rotations in range of full 360 degrees with 10 bit turning resolution. Full turn takes approximately 1.2s. The gimbal itself and possible future external devices are able to communicate via extended VISCA protocol and RS-232C standard. Performed tests and evaluations have shown promising results regardless the time of the day. The presented gimbal shows a great potential for target detection and tracking using UAVs and developing more sophisticated algorithms for stabilization.

In the future work, we will include multiple objects tracking and implement visible light and thermal cameras fusion in order to develop more accurate detection algorithms. Next versions of the gimbal will also utilize the mounted IMU unit for better stabilization and gimbal state visualization. Building upon acquired experience in laboratory ,we will test next implementations during test flights. There is also possibility of implementation of parallel processing algorithms in order to improve overall performance.

References

1. Orejas, M.E.: UAV Stabilized Platform, Master thesis, Department of Control Engineering Faculty of Electrical Engineering, Czech Technical University, Czech Republic (2007)
2. Ying-Chen, D., Yun-Ping, C., Ying-Ying, C., Yan, C.: Survey on image mosaic algorithm of unmanned aerial vehicle. Journal of Computer Applications 31(1), 170–174 (2011)

3. Ji, W., Li, Q., Xu, B., Zhao, D., Fang, S.: Adaptive fuzzy PID composite control with hysteresis-band switching for line of sight stabilization servo system. Aerospace Science and Technology 15(1), 25–32 (2011)
4. Chao, H., Cao, Y., Chen, T.: Autopilots for small unmanned aerial vehicles: A survey. International Journal of Control, Automation and Systems 8(1) (2010)
5. Lee, D., Kaminer, I., Dobrokhodov, V., Jones, K.: Autonomous feature following for visual surveillance using a small aerial vehicles with gimbaled camera system. International Journal of Control, Automation and Systems, Special Section on Advances in Intelligent Visual Surveillance Systems (2010)
6. Sward, R.: Georegistration of imagery from unmanned aircraft systems using Ada. In: Proceedings of the ACM SIGda 2009 Annual International Conference on Ada and Related Technologies, Monterey, CA, USA (2009)
7. Siemieński, P., Denisiuk, P.: Production of a short series of cases by application of printed master pattern and silicone moulds. VIII Forum Inżynierskie ProCAx, Czasopismo Mechanik R. 83 (1), 66–72 (2010)
8. Bieszczad, G., Sosnowski, T., Dabrowski, M.: Method of detection, localization and tracking obejcts with thermovision camera. Elektronika: Konstrukcje, Technologie, Zastosowania 51(1), 84–87 (2010)
9. Biszczad, G., Sosnowski, T., Orżanowski, T., Kastek, M.: Hardware implementation of tracking algorithm on thermovision images. FPGA, Pomiary, Automatyk, Kontrola 55(8), 654–656 (2009)
10. Peteri, R., Siler, O.: Object Tracking using Joint Visible and Thermal Infrared Video Sequences. Laboratoire de Mathématiques Image et Applications. Université de La Rochelle, Cedex (2009)
11. Sony Corporation: SNC-RZ25 VISCA command manual, ver. 1.1 (2005)

The Comparison of Capabilities of Low Light Camera, Thermal Imaging Camera and Depth Map Camera for Night Time Surveillance Applications

Karol Jędrasiak and Aleksander Nawrat

Abstract. Night time surveillance is an important task for existing monitoring solutions. There is a need for low cost and high light-insensitivity solution. In this paper we present a results acquired during night indoor surveillance using three common vision devices and RGB-D camera. Experiments were performed in full light, minimal light and pulsating light simulating light alarm. Additionally 20 people were asked to recognize person in the selected frames of the acquired video sequences. Comparison of detection results and questionnaires answer charts are presented. It is presented that RGB-D cameras show great potential for low cost constant autonomous indoor surveillance regardless of the light conditions in the room.

Keywords: camera comparison, low-light camera, thermal imaging, RGB-D camera.

1 Introduction

Night time surveillance is an important task for existing monitoring solutions. Minimal light amount received by video cameras detectors and low signal to noise ratio (SNR) makes it difficult to detect and identify burglars. Existing solutions are usually based on visible light video cameras which are suitable for usage during day however they experience problems during twilight and are useless during night. Using additional artificial lights during night in order to improve vision conditions is the most basic solution, however it is a costly solution and not commonly used. Recently, low-light cameras are getting more popular. Low-light camera is described as a video camera capable of observing the scene characterized by a illumination

Karol Jędrasiak · Aleksander Nawrat
Silesian University of Technology, Institute of Automatic Control,
Akademicka 16, 44-101 Gliwice, Poland
e-mail: {Karol.Jedrasiak,Aleksander.Nawrat}@polsl.pl

A. Nawrat et al. (Eds.): Advanced Technologies for Intelligent Systems, SCI 440, pp. 117–128.
springerlink.com

lower than 1 lx. Night time capabilities of most of low-light cameras can be further improved by using IR (infrared) illuminators.

For surveillance of strategic buildings and infrastructure, thermal imaging cameras are used. Using thermal cameras is the most costly solution but at the same time the most reliable. Thermovision detectors are unaffected by visible light conditions, therefore they work exactly the same regardless of the day/night and weather conditions.

There is a need for low cost and high light-insensitivity solution. Recently more attention are getting RGB-D cameras, where D stands for depth map acquired from e.g. structural light analysis. Cheap IR projectors can be used to create a video camera suitable for usage in short range but unaffected by light conditions.

There is a need to compare the mentioned video cameras in terms of intruder detection quality during night. In this paper we present a results acquired during night indoor surveillance using four mentioned vision devices. As a quality measurement number of positively detected to all detected frames ratio was assumed. Additionally we decided to ask 20 people interested in computer vision whether they can recognize person in the acquired frames. Results of experiments and questionnaires are presented.

2 Night Time Surveillance Applications

Most of surveillance solutions are based on visible light cameras. It is difficult to perform traditional background subtraction methods for motion and object detection due to small size and low contrast of objects of interest. Using of contrast change instead of pixel intensity change for object detection was presented in [1]. False detections were suppressed using motion prediction and spatial nearest neighbor data association.

Alternative approach is to previously filter night time images in order to use traditional background subtraction methods [2]. A method exploiting the fact that most of surveillance systems continuously operative is presented in work [3]. Night time images are enhanced using day time luminance and reflectance. In paper [4] authors suggested use of independent component analysis (ICA) for indoor motion detection using background subtraction. First the background image is used to train ICA model, then the ICA is used to separate foreground and background images.

Some solutions use more than single camera in order to increase the quality of the detection. In paper [5] a stereo imaging system of low-light cameras and IR illuminators was used to detect and measure craniofacial anthropometric measurements with an accuracy of 27 mm within 3m range. Strong shadows and light reflections during night time conditions are disturbing motion detection. Shadow removal algorithm used in [6] is based on an assumption that equation (1) is fulfilled. The DELTA value was experimentally set as 0.1.

$$max = \left\{ \left| \frac{R_{fr}}{R_{bg}} - \frac{G_{fr}}{G_{bg}} \right|, \left| \frac{G_{fr}}{G_{bg}} - \frac{B_{fr}}{B_{bg}} \right|, \left| \frac{B_{fr}}{B_{bg}} - \frac{R_{fr}}{R_{bg}} \right| \right\} \leq \delta, \tag{1}$$

where:

R_{fr}, G_{fr}, B_{fr}, current frame values,

R_{bg}, G_{bg}, B_{bg}, background model values.

There is also a wide research area of using near-infrared cameras and thermal imaging cameras for night time surveillance. Both types of cameras can operate in conditions which human eye perceives as a total darkness. In paper [7] it is presented that traditional motion detection algorithms work well on illumination invariant images acquired from the mentioned video devices. Work described in [8] shows that simple calibrating thermal imaging camera for range of temperatures 30C-40C is sufficient for object detection using simple thresholding therefore easily done in real-time. Similar approach was used in home alone faint detection [9].

Thermal imaging cameras field of view (FOV) is relatively small (in range 10-20 degrees) in comparison to visible light cameras. To surpass this limitation in paper [10] omnidirectional system was presented. Wide FOV was acquired using hyperbolic IR mirror.

3 Vision Devices

Usually visual surveillance is based on video cameras. Images acquired using video cameras are two dimensional matrices composed of sensor responses that can be described by (2) as presented in [11]:

$$p^x = e^x \cdot n^x \int_{\omega} S^x(\lambda) E(\lambda) R(\lambda) d\lambda, \tag{2}$$

where:

p^x,3-vector of sensor responses,

S^x, surface reflectance,

E, spectral power distribution of the illumination,

R, 3-vector of sensitivity functions of the video camera,

ω,visible spectrum range

e^x,unit vector in direction of the light source,

n^x,unit vector corresponding to the surface normal at x,

Video device is therefore every device which output signal is also a two dimensional matrix of 3-vector responses.

The most popular vision device is a light video camera. For experiments we used a Logitech QuickCam camera (fig. 1). Its equipped with CCD sensor and its output format consists of one byte per R, G, B channel. Image resolution is 320x240 pixels. Video frame rate is 25 frames per second. Its parameters are typical for webcam cameras.

The low-light camera we used was a Sony IPELA SNC-RZ50N model (fig. 2). It is a PTZ camera with a maximum image resolution of 711x485 pixels and night-

Fig. 1 A Logitech QuickCam webcam

Fig. 2 A Sony IPELA SNC-RZ50N IP camera

mode function. Using night-mode it is possible to acquire clear images in low-light environments around 0.3 lx. Depending of the current zoom level camera's FOV is in range from 1.7 to 42 degrees.

The long-wave thermal imaging camera used was Thermoteknix 110K model (fig. 3). Its detector resolution is 284x288 and frame rate can be as high as 50 frames per second. Detector pitch is $25\mu m$ and spectral response is in range $8 - 12\mu m$. As a thermovision camera it is visible light invariant therefore it is best suited for continuous day/night time surveillance applications. Unfortunately its FOV is only 15.6 x 11.7 degrees.

RGB-D are video devices which output is depth map of the observed scene. At the time of research there were three main methods of acquiring depth maps. First and the most popular was to use stereo vision of two calibrated cameras. Second method is counting length of flight of emitted photons. This method is used by time-of-flight cameras. Recently a lot of attention received generating depth maps from structural light because of Microsoft Kinect device (fig. 4).

We have measured that Kinect is using near-infrared 830 nm (fig. 5) illuminator in order to display a known pattern onto the scene and a IR camera is used to calculate the depth. IR camera FOV is 57x43 degrees. It is believed [12] that depth is computed using triangulation against a known pattern from the IR projector. For each pixel in the acquired image a correlation window is used to compare the local

Fig. 3 A Thermoteknix 110K thermal imaging camera

Fig. 4 A Microsoft Kinect game controller (RGB-D camera)

pattern with the pattern at the known depth. It is used to compute disparity and the depth is computed using equation that comes from stereo vision systems (3):

$$z = \frac{b \cdot f}{d}, \tag{3}$$

where:

z, depth at the pixel point with disparity d,

b, baseline between stereo vision system cameras,

f, common focal length.

Depth data acquired from RGB-D camera can be used in fusion with RGB camera in order to perform real time segmentation and tracking of objects in 3D space rather than its 2D projection onto the camera lenses [13]. It also possible to build rich 3D indoor maps using RGB-D mappings [14].

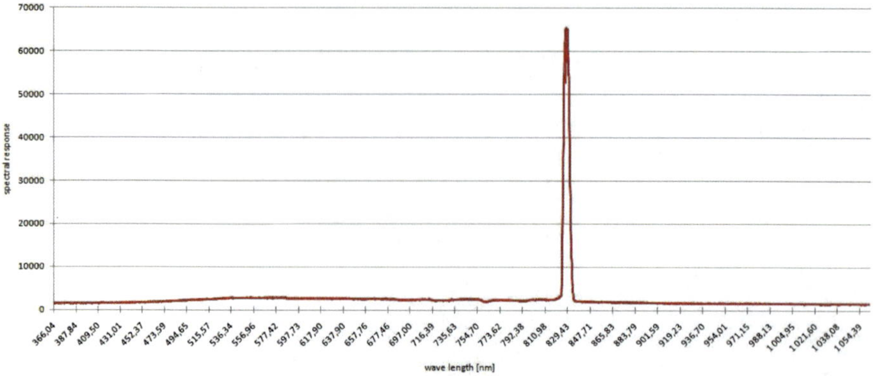

Fig. 5 Spectral response of the Kinect's IR projector

There are also other vision devices but they are not so well researched as the ones presented above. One of the most interesting are the 3D scanners based on radio frequencies echoes.

4 Experiments Description

Goal of the tests was to compare potential of tested devices for indoor surveillance during night. Experiments were divided into two parts. First one tested how easy information from the cameras can be used by standard motion detection algorithms for autonomous object detection. Second test's goal was to choose which camera is best in potential monitoring operator for night time surveillance.

4.1 Motion Detection by a Computer

During part one, three partial experiments were performed. For all three experiments the observed scene remained the same (fig.6). In the left side of the image there are multiple windows. On the right side, there is a door to the inner corridor of the building. In the background, various items and a wall can be seen. Actor's task was to jump in through the window in the left side of the scene then sneak around to the door in the right side of the room. All four video devices streams were recorded at the same time using developed video recording application.

As a quality factor a ratio r (4) computed as ratio of all frames with a person to detected person frames was used:

$$r = \frac{\sum_{i=0}^{n} I(n)}{n}, \tag{4}$$

Fig. 6 Scene used for the experiments

where:

n, number of all detected frames by motion detection algorithm,

$I(n)$, binary function that takes as input index of the frame and return 0 or 1 depending if the frame was tagged as a detected.

As a motion detection algorithm for visible and RGB-D cameras we used standard background subtraction algorithm with background image modeled as follows (5):

$$B_{t+1} = \alpha I_t + (1 - \alpha) B_t, \quad (5)$$

where: B_{t+1}, segment background value,

I, segment median intensity,

α, speed of background update.

For RGB-D cameras additionally empty black pixels were omitted during background updating.

However, as suggested in [6] for thermal camera, it was decided to use a different motion detection algorithm. The global image average as a reference value for thresholding. Each pixel with intensity value different then average background value by beta parameter was assumed as a foreground pixel. Beta parameter was experimentally set as 0.3.

First experiment was performed with the light on as a training run for the motion detection algorithms and the actor. Sample frames from the trial were used for questionnaire. Second experiment was performed during night with the lights off. In almost complete darkness actor's task was to jump in through the window, sneak through the whole scene and escape via the door in the right side of the observed scene. Finally during the third experiment the flashing light was used. It was an imitation of a visible light alarm that could be equipped in e.g. stores. During all trails actor performed the same task only the light conditions were different.

4.2 Motion Detection by a Human

Motion detection in indoor nighttime surveillance applications is usually performed by a trained human operator. In order to compare the results of automatic motion detection and a natural human ability to detect other humans in static images from different vision devices we created a questionnaire. Twenty potential monitoring operators were asked to answer simple questions whether they can see a person and if they can distinguish gender of the person. Images used in questionnaire are presented in fig. 7. Acquired results are presented and compared with the automatic results in the following section.

Fig. 7 Images from the four vision devices used during experiments that were chosen in order to compare automatic and manual results using questionnaire

5 Results

After the first experiment performed in the full light conditions it was no surprise that all video cameras scored high result. The visible light camera's score was 0.959. The ratio for low light camera was 0.957 and for thermovision 0.967. Bright scene, no shadows and camera movement are optimal conditions for background subtraction algorithm. RGB-D camera scored a result 0.961. It is a value comparable with traditional video devices. Results of the corresponding question from the questionnaire are presented in fig. 8. It can be seen that potential monitoring operators were able to easily fulfill the task of finding person in the images from the traditional video devices. RGB-D cameras produce a scene visualization that is different type than video devices, therefore respondents were slightly confused and their answers were less confident.

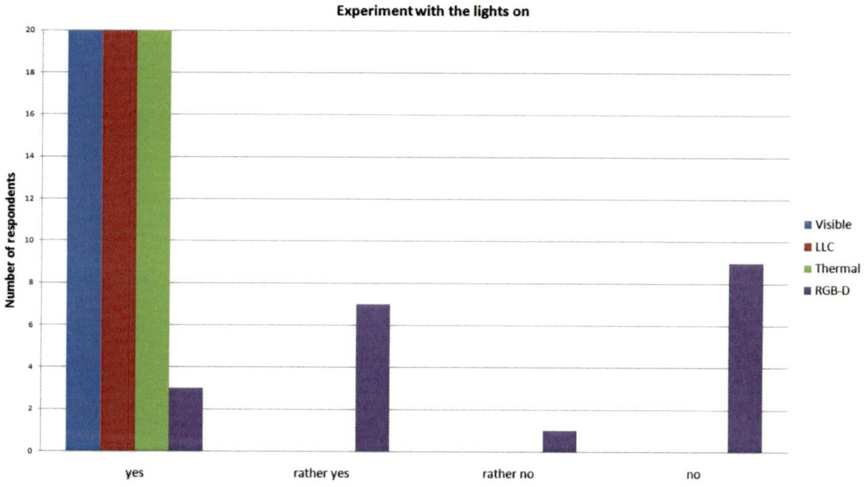

Fig. 8 Results of the questionnaire's part related to the first experiment

Experiment number two was performed in the conditions perceived by a human as almost complete darkness. Therefore images acquired from visible light camera contains only noise and result ratio is equal to 0. However in images from low light camera motion detection algorithm was easily able to detect person. The ratio was 0.903 which is comparable result with the full light experiments. There was no difference in images acquired using thermal imaging and RGB-D cameras. Human respondents also were not able to detect person in images from visible light camera. Detailed results are presented in fig. 9.

Experiment with the pulsating light imitating light alarms often installed in stores acquired interesting results. Human respondents' answers (fig. 10) were comparable with those of the previous experiment. However video devices results were affected. Visible light camera's results in darkness was the same. Pulsating light affects all pixels intensities that are visible in the images from low light cameras therefore successful detection occurs only in the interval between pulses. The result ratio was measured as 0.531 while thermovision and RGB-D cameras proved to be illumination invariant and therefore optimal for indoor surveillance. Their results were respectively 0.981 and 0.983.

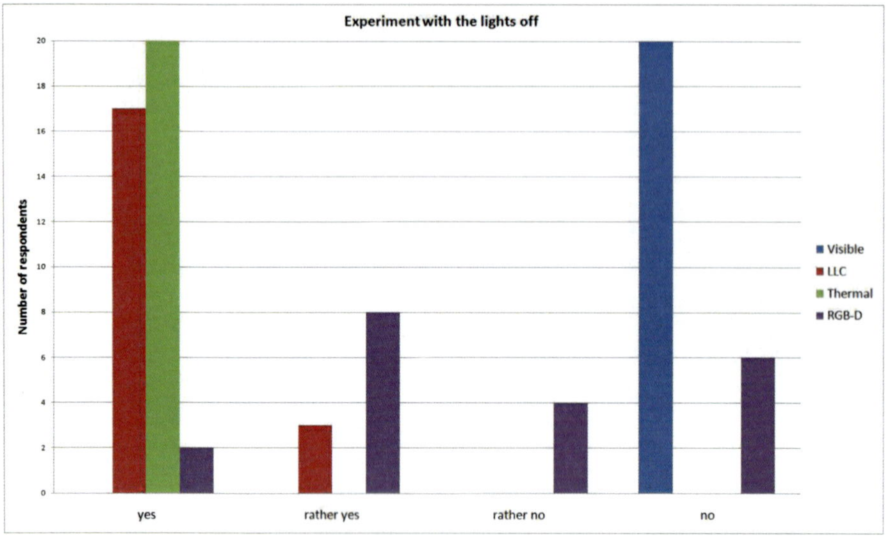

Fig. 9 Results of the questionnaire's part related to the second experiment

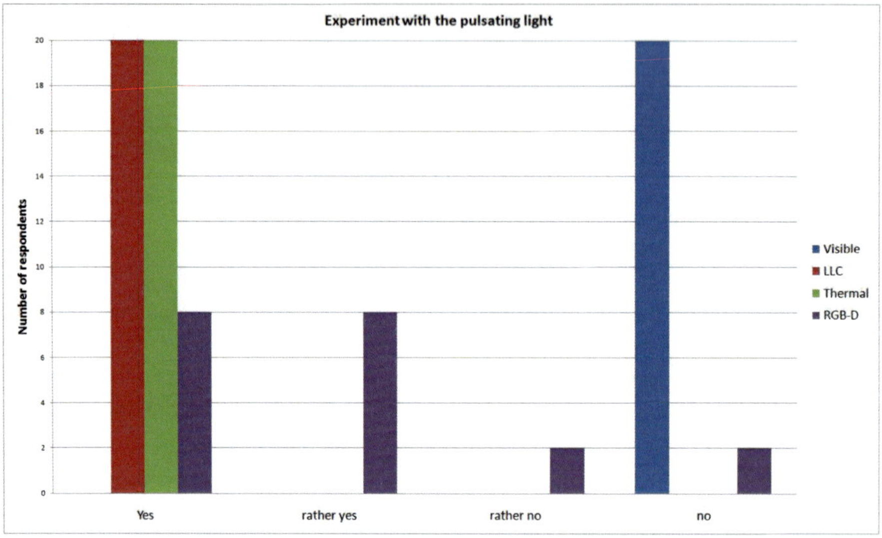

Fig. 10 Results of the questionnaire's part related to the third experiment

6 Conclusions

In this paper the three traditional surveillance video devices were tested against an alternative solution based on RGB-D camera. Capabilities for indoor day/night surveillance applications were tested during three experiments and their goal was to detect person in the acquired image sequences from the camera. Additionally 20 people were asked to answer simple questions about images acquired during the experiments. Human and traditional motion detection by background subtraction algorithms capabilities for object detection were compared and the results are presented. Acquired results shows that humans are not used to different type of scene visualization. Vision devices are compared using a quality ratio computed as a successfully detected frames to all possible for detection frames ratio. It is shown that both expensive thermovision camera and very low cost RGB-D camera are illumination invariant and ideal for indoor surveillance during day and night. It was also shown that existing solutions based on LLC are vulnerable for often installed close light alarms. Additional RGB-D camera feature is that experiments were based only on two dimensional scene projection. Acquired depth map makes it possible to use alternative motion detection algorithms in three dimensional space in order to significantly increase detection quality. Also anthropomorphical measurements could be performed for e.g. person identification.

In the future, RGB-D cameras capabilities will be tested in large halls and outdoor in order to precisely measure system limitations. Alternative way of research is developing motion detection and recognition algorithms that utilize three dimensional depth data acquired from low cost RGB-D cameras.

References

1. Huang, K., Wang, L., Tan, T., Maybank, S.: A real-time object detecting and tracking system for outdoor night surveillance. Pattern Recognition 41, 432–444 (2008)
2. Wang, Y., Fan, C.: Moving Object Detection for Night Surveillance. In: Proceedings of 2010 Sixth International Conference on Intelligent Information Hiding and Multimedia Signal Processing, pp. 236–239 (2010)
3. Yamasaki, A., Takauji, H., Kaneko, S., Kanade, T., Ohki, H.: Denighting: Enhancement of nighttime images for a surveillance camera. In: 19th International Conference on Pattern Recognition, Sapporo, pp. 1–4 (2008)
4. Tsai, D., Lai, S.: Independent Component Analysis-Based Background Subtraction for Indoor Surveillance. IEEE Transactions on Image Processing 18(1), 158–167 (2009)
5. Chong, A., Ariff, M., Majid, Z., Setan, J.: Night-time surveillance system for forensic 3D mapping. In: 2010 3rd International Congress on Image and Signal Processing, pp. 502–506 (2010)
6. Gupta, A., Ashis Pati, K., Venktesh, K.: Indoor Multi-Camera Surveillance System, Mid Term Project Report. Indian Institute of Technology, Kanpur (2009)
7. Ariff, M., Majid, Z., Chong, A.: Near-Infrared Camera For Night Surveillance Applications. Geoinformation Science Journal 10(1), 38–48 (2010)
8. Wong, W., Tan, P., Loo, C., Lim, W.: An Effective Surveillance System Using Thermal Camera. In: Proceedings of International Conference on Signal Acquisition and Processing, ICSAP 2009, Kuala Lumpur, Malaysia, pp. 13–17 (2009)

9. Wong, W., Lim, H., Loo, C., Lim, W.: Home Alone Faint Detection Surveillance System Using Thermal Camera. In: Proceedings of the 2010 Second International Conference on Computer Research and Development, Kuala Lumpur, pp. 747–751 (2010)
10. Wong, W., Tan, P., Loo, C., Lim, W.: Omnidirectional Surveillance System Using Thermal Camera. Journal of Computer Science and Engineering 3(2), 42–51 (2010)
11. Schaefer, G.: How useful are colour invariants for image retrieval? Computer Vision and Graphics, Computational Imaging and Vision 32, 381–386 (2006)
12. Internet, `http://www.ros.org` (visited June 1, 2011)
13. Bleiweiss, A., Werman, M.: Fusing Time-of-Flight Depth and Color for Real-Time Segmentation and Tracking. In: Kolb, A., Koch, R. (eds.) Dyn3D 2009. LNCS, vol. 5742, pp. 58–69. Springer, Heidelberg (2009)
14. Henry, P., Krainin, M., Herbst, E., Ren, X., Fox, D.: RGB-D Mapping: Using Depth Cameras for Dense 3D Modeling of Indoor Environments. In: Proceedings of the 12th International Symposium on Experimental Robotics, Delhi, India (2010)

Object Detection Using IR Camera

Aleksander Nawrat, Krzysztof Daniec, and Tomasz Warmuz

Abstract. The main goal of this paper is to implement and test various target detection algorithms in thermal imagery. The topic of target detection and moving target detection in infrared video streams is considered as a very challenging research topic in computer vision discipline. Although there is a huge number of different algorithms developed for target detection and target tracking in video streams generated by daylight cameras, there is still a limited number of solutions in field of infrared video streams.

This document introduces some of the basic techniques for image processing and some either classical approaches or state-of-the-art algorithms for target detection in IR imagery. Some of the chosen algorithms are implemented, tested and described in more details.

Keywords: object detection, IR video, blob detection, labeling.

1 Problem Description

The main problem faced during implementation of a target detection algorithm is such that computers perceives images differently than humans do. Human is able to perceive an image or at least a part of it at once and can immediately draw some conclusions about objects placement or target movement. On the opposite, computers, and all computer algorithms for image processing are working in pixel-by-pixel manner, thus each wrong assumption can cause bad results and wrong conclusions can be drawn. Therefore it is practically impossible to design a robust and 100%

Aleksander Nawrat · Krzysztof Daniec · Tobiasz Warmuz
Silesian University of Technology, Institute of Automatic Control,
Akademicka 16, 44-101 Gliwice, Poland
e-mail: {Aleksander.Nawrat,Krzysztof.Daniec}@polsl.pl,
tobiasz.warmuz@gmail.com

A. Nawrat et al. (Eds.): Advanced Technologies for Intelligent Systems, SCI 440, pp. 129–142.
springerlink.com

error-proof algorithm of target detection and even there are some relatively good and robust solutions, they have no chances while competing with a human vision system.

Thermal images characterizes with a significantly lower signal-to-noise ratio comparing to ordinary daylight images. This means that the power of the noise in the IR images is very high comparing to the power of the signal (information included in the image).

Huge advantage of thermal images is that even when the scene is poorly illuminated or not illuminated at all, thermal detector can sense the target due to the difference of the amount of thermal radiation. Thus it is the easiest case of tracking a mammal in a natural environment where the target (mammal) is much brighter on the thermal image comparing to the background due to its higher body temperature. This assumption was taken in [1, 2]. Unfortunately it is not always a good solution to follow with this assumption because there can be several real-life situations of tracking an object much colder that the background is. For example one can imagine a case of a highway on a warm and sunny day and the task is to track the air-conditioned cars driving on that motorway. Additional problem arises when we consider an ego-motion of the sensor, when the camera is mounted on a moving platform like in airborne IR imagery. In such cases the whole image scene moves either the targets or the background. In order to distinguish the background and the targets some motion compensation algorithms are needed. Such an algorithm for motion compensation needs to assume that the direction of the camera is pointed relatively perpendicularly to the scene because otherwise, when the scene includes more perspective, some of the further slowly moving targets may be not detected due to higher movement of the background on the front of the scene. According to the assumption of non-stationary sensor platform the algorithm needs to be additionally robust against basic transformations such as rotation, translation and scaling. The solution presented in [3] assumes a stationary sensor platform. Moreover, [3, 5, 6] provides also some solutions that assumes that the target feature do not change drastically over the course of tracking.

This kind of obstacles needs to be recognized at the very beginning stage of the algorithm implementation and all of the possible assumptions needs to be stated. This shows that algorithms for target detection may significantly differ in different applications and at the same time the wider the application we assume to be, the lower robustness of the algorithm we will get.

Nevertheless the ideal algorithm needs to work properly in different conditions it also needs to be computationally light and simple so it can work in real time and the result with a relatively smooth video stream (at least few frames per second).

2 Algorithm Decription

Recalling the assumptions for the algorithm stated in section 1 we know that the potential targets are assumed to be a regions of brighter pixels comparing to the background. In order to enhance or to mark this kind of regions an algorithm called blob detection will be used.

2.1 *Basic Idea*

Blob detection results with a binary image that indicates the regions of brighter or darker pixel intensities comparing to the background. According to our assumptions we will focus on brighter regions and skip the case of darker regions. On this binary image, white regions indicates blobs. Using blob detection algorithm one can easily distinguish the objects and separate them from the background. In order to distinguish between different blobs another algorithm needs to be used which is blob labeling. This algorithm is responsible for proper labeling of blobs where each separate blob gets its own unique label. Finally having all blobs detected from the scene and properly labeled it is possible to get some characteristic features of each object like area, circularity factor, the mass center or others. Having the area for each blob known, one can also discard relatively small blobs that most likely results from inaccuracy of the IR detector.

Fig. 1 Visualization of blob detection and blob labeling algorithm

The fig. 1 presents the simplified scheme of operation and the basic idea of the blob detection and blob labeling algorithm. The first image to the left presents a sample grayscale IR image with a scene of two persons crossing the road. The algorithm of blob detection results with two blobs indicating the shape and position of the targets on the scene. The result of the blob detection algorithm is illustrated by the middle image. Finally the most right image illustrates the result of the blob labeling algorithm where both blobs gets their unique labels. Colors are used only for better visualization that both blobs are treated separately as independent blobs.

2.2 *Algorithm Decription*

As it was mentioned before, blob detection algorithms are focused on detection of brighter regions of image. The algorithm uses the fact that each flat region is separated from the background with a steep edge. Looking at the fig. 1 one could think of an algorithm for blob detection which uses simple thresholding method but unfortunately it will not work correctly in case when the background is more significantly nonuniform. Apart from that, thresholding methods are very sensitive for any changes of background or object intensity. Fig. 2 presents results of applying thresholding method on sample IR image. Such an algorithm may give us satisfactory results in case of single image but when we consider a video stream, where the scene changes and objects moves it would be necessary to adapt threshold value according to changing conditions. The problem is what would that threshold depend on? The image on the left from the fig. 2 corresponds to the threshold value 100 while the right image corresponds to the threshold value 150.

Fig. 2 Results of thresholding method

In section 1 it was assumed that the algorithm needs to deal with the nonuniform background and deal with changes of background and object intensity. The fig. 3 presents a situation where it is impossible to determine such a threshold value to make the thresholding method work for detecting each of the three objects from the scene.

That situation, mentioned above, motivates the usage of edge detectors in blob algorithms. Edge detection algorithms are focused on detection of image regions where there are drastic changes in luminance between relatively smooth regions while thresholding methods are only focused on each pixel luminance separately.

Fig. 3 Phantom image with three objects on the nonuniform background

3 Blob Detection

The results from edge detection algorithms motivates the idea of introducing more smoothing LoG filter comparing to the popular one described on [7]. Let us increase both parameters of LoG filter i.e. the size of the window and σ in order to gain a smooth and noise-free image. Let us introduce on a fig. 4 a 19×19 large filter window with $\sigma = 2.5$.

Fig. 5 presents the result of applying the modified LoG filter and due to those changes of LoG filter, the mid-cross-section of the resulting image from fig. 5 presented on fig. 6 has different (smoother) shape comparing to the result from edge detection algorithm. Yet again, red dots points the position of the edge of the object.

Recalling the assumptions for the algorithm stated in section 1 we know that the potential targets are assumed to be a bright regions (due to their higher temperature comparing to the background) observed from above (the IR camera is mounted on UAV). According to those assumptions we know that potential targets are most likely small bright regions on darker background as it can be seen on fig. 1. Since the potential targets are assumed to be relatively small it can be enough to apply a threshold method to get the inner part of the bright object.

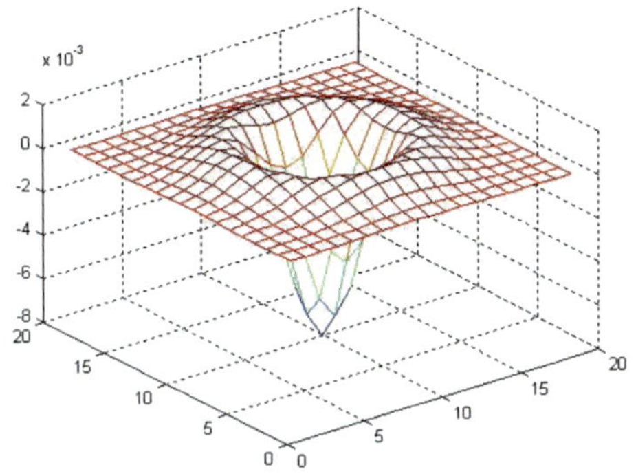

Fig. 4 The 19×19 version of LoG filter with $\sigma = 2.5$

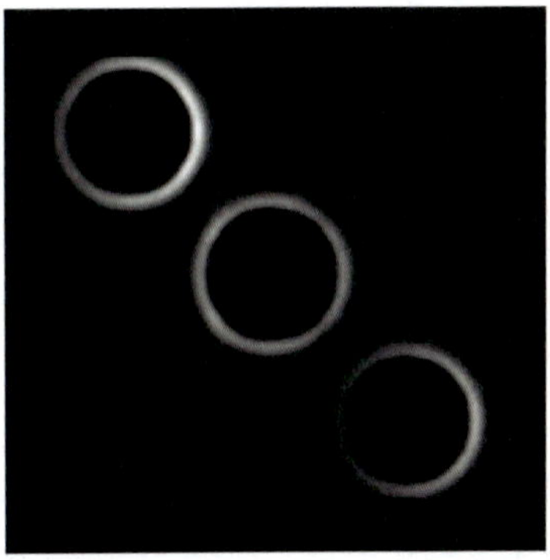

Fig. 5 Result of applying the LoG filter

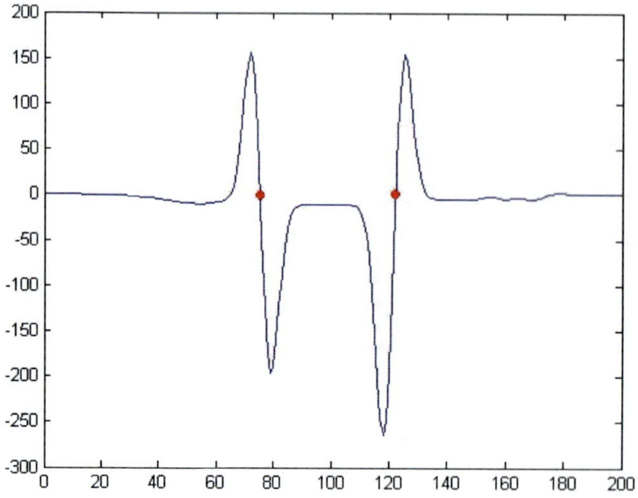

Fig. 6 The cross-section of the smoothed LoG

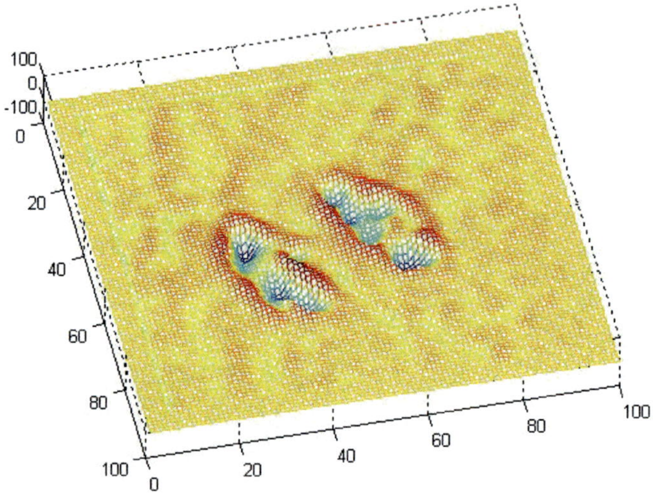

Fig. 7 The 3D visualization of smoothed LoG applied on real IR image

On the fig. 7 the result of applying LoG filter is presented. One can see that regions that corresponds to position of objects (brighter regions) has negative values (on the figure it can be seen as blue regions) due to occurrence of object edges. This is why for blob detection it may be sufficient to detect regions with negative values of smoothed LoG using thresholding method. The fig. 8 presents the sample real IR image on the left-hand-side, the image filtered with LoG in the middle and on the right-hand-side the resulting binary image $B(x,y)$ applying thresholding method (white regions corresponds to regions of negative values). The resulting binary image shows two significant (the biggest) blobs that corresponds to the desired objects and a number of redundant small blobs.

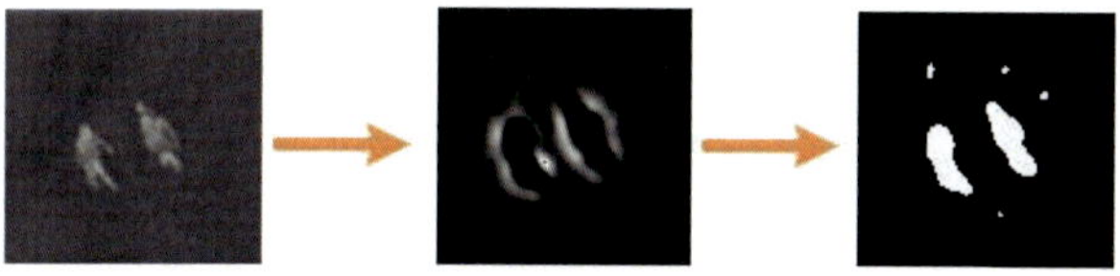

Fig. 8 Results of blob detection algorithm applied on sample real IR image

The results seems to be satisfactory and takes us closer to the desired results as it was presented on fig. 1. Thus the next stage is to filter the binary image to discard small redundant blobs. In order to do this it is necessary to attach a unique label for each blob, then to calculate the area for each of blobs and next discard potentially non-significant blobs with area smaller than some value. Next section describes method for blob labeling.

4 Blob Labeling

The algorithm for blob labeling is needed to set a unique label for each separate blob. As a separate blob we understand a blob that has no other blob in 4-neighborhood. As 4-neighborhood of pixel for binary image B at $B(i,j)$ we understand pixels at $B(i-1,j)$, $B(i,j-1)$, $B(i+1,j)$ and $B(i,j+1)$. If in 4-neighborhood of pixel $B(i,j)$ there is a blob pixel it means that it is the same blob thus the same label for the neighboring blob needs to be assigned. There are different algorithms of blob labeling [4]. Fig. 9 presents the basic idea for the so called two-pass algorithm. In the first pass the algorithm goes pixel-wise and is looking for unlabeled blob pixels, where there are no labeled pixels in the neighborhood. This state of the algorithm is represented by dark gray arrows. While the algorithm finds a unlabeled pixel with no labeled pixels in the neighborhood (red arrows) it sets a unique label for the pixel. Finally, blue arrows represents the state of the algorithm when it finds an unlabeled pixel with other labeled pixel in the neighborhood. Then the label is set equal to the labeled neighboring pixel. Here, on the fig. 9 it is shown, that in case of concave objects it may happen that the same blob can get two separate labels (on the figure it is depicted with two different pixel colors). In order to avoid such cases it is necessary to store the information about occurrences of situations where two neighboring pixels has different label. Different labels that corresponds to the same blob are stored in classes and in the second pass of the algorithm each pixel label is replaced with the minimum label from the class. More details for the two-pass algorithm can be found in [4].

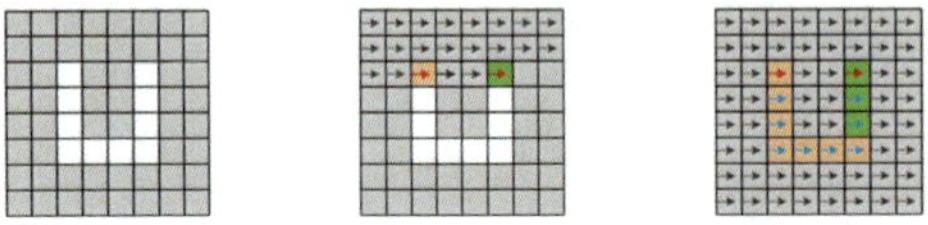

Fig. 9 Idea of the first pass from the two-pass algorithm of blob labeling

There is also another way for blob labeling that do not require to pass through the image more than once. On the fig. 10 one can see the simplified scheme of operation for the one-pass recursive algorithm of blob labeling.

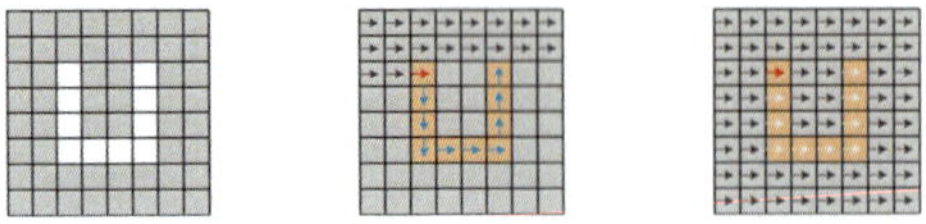

Fig. 10 Idea of the recursive algorithm of blob labeling

Similarly as in the algorithm presented on fig. 9 the dark gray arrows are looking for an unlabeled blob pixel. Once a new unlabeled blob pixel is found (red arrow) a unique label is set to the pixel and a recursive algorithm is ran for setting the same new label for all blob pixels in 4-neighborhood. As long there is a unlabeled blob pixel in the neighborhood the recursive algorithm is running (blue arrows). Finally, when whole blob is labeled with the same label, the algorithm goes back at the position of the first pixel from the blob and starts looking for another unlabeled blob pixel (dark gray arrows). White arrows corresponds to situations where an already labeled pixel is found. Since potential objects are assumed to be small bright regions, the recursive algorithm do not introduce too much burden for the processor. The result of recursive blob labeling algorithm for sample real IR image is presented on fig. 11.

The last stage of blob detection algorithm is to discard small redundant blobs that do not corresponds to desired objects. The algorithm used for that is also a two-pass method. First, we count the area for each blob by counting the number of pixels with the same label and next in the second pass, the pixels with labels that corresponds to small blobs (smaller than some threshold value) are set to zero thus they are

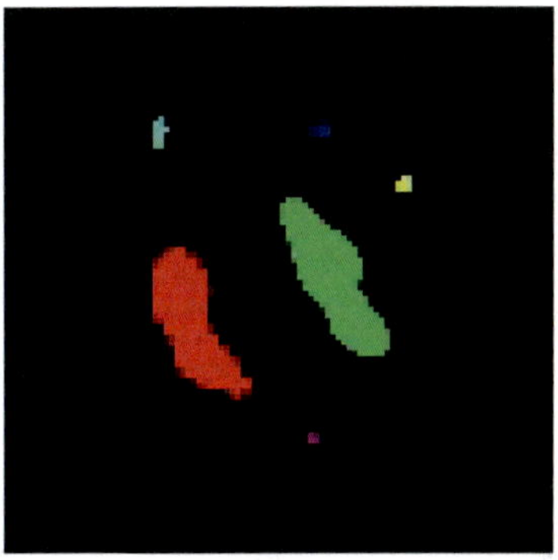

Fig. 11 Result of recursive blob labeling algorithm applied on sample real IR image

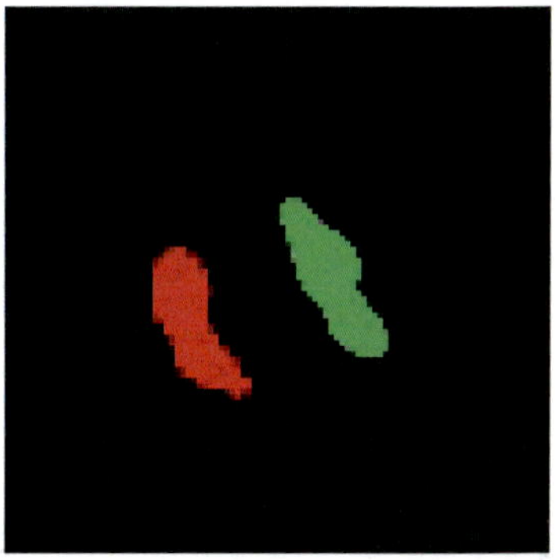

Fig. 12 Result of algorithm for discarding small and redundant blobs

Fig. 13 Result of the blob detection and blob labeling algorithm applied on a real IR image

no longer visible on the output screen. The result of such algorithm is presented on fig. 12.

Finally as a result, we are able to get relatively close to the desired result as presented on fig. 1. The fig. 13 presents the results of blob detection and blob labeling algorithms according to the results from fig. 1 and 8. Looking at the result from the fig. 13 one can notice that the algorithm enabled us to detect the lantern which has a relatively low temperature comparing to the pedestrians. In order to extend the algorithm, one can think of filtering out the low temperature objects in order to enhance the pedestrians. Furthermore, having such blobs, one can think of calculating some characteristic features for each blob i.e. some shape factors or mass center. For example it might be also possible to track the mass centers of each blob for further compensation of steady objects and enhancement of moving objects. Having the trajectory of the objects movement from number of the past frames one may try to apply some predictive algorithms in order to predict future movement of the object.

5 Results of the Algorithm

First fig. 14 presents the result of the blob detection algorithm on the sample IR video where three warmed batteries are rolled on the cold floor. The contrast between the batteries and the floor is very low and the signal-to-noise ratio is very low as well. The batteries are rolling on the floor rather rapidly but it do not cause any problem for the algorithm and each battery is properly detected as separate objects. The detected objects are enhanced by setting the R (red) channel to 255 for each pixel that to corresponds to the certain blob. the red channel is not accidentally chosen because red channel provides less information than green channel but more information than blue channel. Moreover red is better visible on graylevel images comparing to blue. Fig. 15 presents the result of blob detection algorithm on sample IR video. One can distinctly see that each pedestrian is properly detected as separate object and additionally some other objects has been detected i.e lanterns and road signs. Additionally fig. 16 presents the same scene as fig. 15 but shifted. The pedestrians are displaced and new object has appeared on the scene namely the car. On the bottom right corner of both frames there is a region of warmer ground due to some underground municipal waste pipings. Because the contrast of those piping is lower and the edges are not that sharp comparing to objects like pedestrians or cars the blobs are rather weak and small.

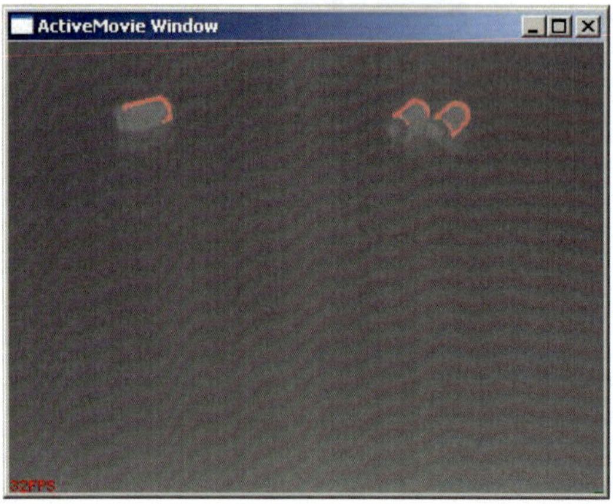

Fig. 14 Result of the blob detection algorithm on the sample IR video

Next two fig. 17 and 18 presents the another set of results for blob detection algorithm on sample IR video. On the scene there are several pedestrians and cars. Pedestrians are much warmer comparing to the cars thus they are much better detected by the algorithm. The video has been captured in winter thus it can be the reason that the cars are very cold and it is difficult for the algorithm to detect them. On the scene there are also several windows detected. The situation gets more

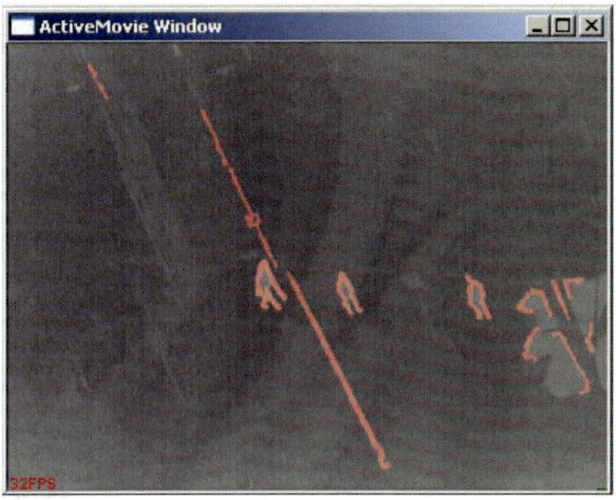

Fig. 15 Result of the blob detection algorithm on the sample IR video

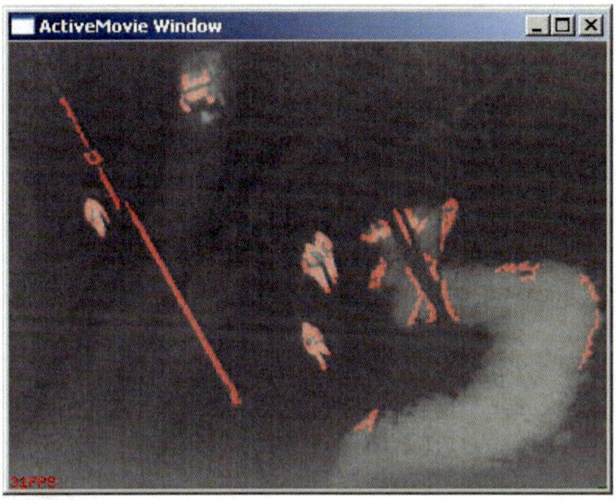

Fig. 16 Result of the blob detection algorithm on the sample IR video

complicated if the scene on the IR video do not follow the second assumption stated in section 1 that the potential target is supposed to be much brighter comparing to the background. This situation happens very often in hot summer time on the street. One can imagine very hot asphalt street and pedestrians that are colder than the street what results in the scene where potential targets (pedestrians) are darker than the background (street). Cars are also darker than the street due to the fact that they use air conditioning very often. Results of such unfortunate case is presented on fig. 19 and 20. Fig. 19 presents the scene where the car and one pedestrian are in a shadow made by the tree nearby. Due to the very high sunlight illumination there

Fig. 17 Result of the blob detection algorithm on the sample IR video

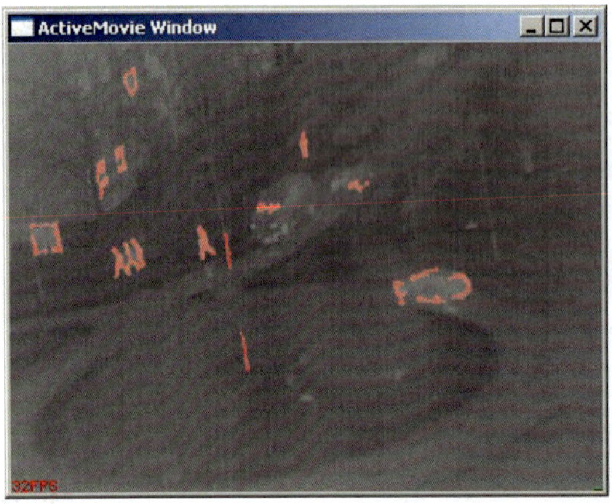

Fig. 18 Result of the blob detection algorithm on the sample IR video

are plenty of false and weak blobs. Either the car or the pedestrian are detected properly. Next fig. 20 emphasizes the situation where the background is brighter comparing to the potential target (against second assumption stated in section 1) In such cases the blob regions are marked on the other side of the desired object what is obviously improper. One can see this effect on the bottom of the car on the fig. 20. The results would be more likely proper if the videos would be captured during the night when there is no influence of the strong sunlight radiation. This considerations shows that the algorithm results strongly depends on weather conditions and that it is very difficult to implement an universal algorithm.

Fig. 19 Result of the blob detection algorithm on the sample IR video

Fig. 20 Result of the blob detection algorithm on the sample IR video

6 Conclusions

The results for the algorithms presented in section 2 shows that it is very difficult to design such an algorithm that would provide proper results in various weather conditions, for various kind of potential targets observed from different distances. The presented algorithm for blob detection gives satisfactory results as long as the assumptions stated in section 1 are satisfied. The most problematic case is described in section 2 where the background is brighter comparing to the desired target. In such case, the blob detection algorithm doesn't mark the object but it only enhance its outer contour. Such case may happen in sunny day during summer. It shouldn't be a problem to modify the blob detection algorithm in such way that it would handle such case. Nevertheless for the Harris corner detection algorithm there is no difference whether the object is brighter than the background or not and this is an advantage.

References

1. Chen, J.Y., Reed, I.S.: A detection algorithm for optical targets in clutter. IEEE Transactions on Aerospace and Electronic Systems 23(1), 46–59 (1987)
2. Longmire, M.S., Takken, E.H.: Lms and matched digital lters for optical clutter suppression. Applied Optics 27(6), 1141–1159 (1988)
3. Davies, D., Palmer, P., Mirmehdi: Detection and tracking of very small low contrast objects. In: Proceedings of 9th British Machine Vision Conference, pp. 599–608 (1998)
4. Shapiro, L., Stockman, G.: Computer Vision. Prentice Hall (2002)
5. Strehl, A., Aggarwal, J.K.: Detecting moving objects in airborne forward looking infrared sequences. Machine Vision Applications Journal 11, 267–276 (2000)
6. Choudhary, M., Braga-Neto, U., Goutsias, J.: Automatic target detection and tracking in forward-looking infrared image sequences using morphological connected operators. Journal of Electronic Imaging 13(4), 802–813 (2004)
7. Wang, R.: Sharpening and Edge Detection (September 2009), `http://fourier.eng.hmc.edu/e161/lectures/gradient/gradient.html`

Characterization of Hokuyo UTM-30LX Laser Range Finder for an Autonomous Mobile Robot

Piotr Demski, Michał Mikulski, and Roman Koteras

Abstract. One of the most common problems of autonomous mobile robots is object avoidance in a dynamically changing environment. The effectiveness of algorithms responsible for trajectory planing is largely dependent on the correct sensory input. It is necessary to equip a mobile robot with the proper sensors to allow a correct functioning in unknown terrain. With the high degree of complexity standard point distance sensors are insufficient for high speed movement. The recent development of compact laser range finders allowed the minimization of robot dimensions. Hokuyo UTM-30LX is an example of such a sensor. For a precise understanding of the measurements, a characterization of a sensor is needed. This paper is summarizes the parameters of Hokuyo UTM-30LX laser range finder, in particular: drift effect, influence of target distance, surface brightness, color and material, and the sensor orientation. The parameters measured prove that the Hokuyo UTM-30LX can be used in a mobile robot system for complex object detection and avoidance.

Keywords: laser range finder, Hokuyo UTM-30LX, autonomous mobile robot, object avoidance.

1 Introduction

The constant increase in computing power of microcontrollers is allowing mobile robots to become more and more intelligent. With the higher degree of complexity in the control algorithms a certain level of autonomy can be achieved. One of the general, and yet the most profound functions of any mobile robot is movement.

Piotr Demski · MichałA. Mikulski · Roman Koteras
Silesian University of Technology, Institute of Automatic Control,
Akademicka 16, 44-101 Gliwice, Poland
e-mail: piotr.demski@gmail.com,
{Michal.Mikulski,Roman.Koteras}@polsl.pl

A. Nawrat et al. (Eds.): Advanced Technologies for Intelligent Systems, SCI 440, pp. 143–153.
springerlink.com

It consists of several problems depending on the type of the robot. Wheeled mobile robots, humanoids or even UAVs face the problem of object avoidance. Although the problem can be simplified for a preprogrammed environment, missions in real, dynamically changing terrain require a higher degree of knowledge of the surroundings.

The effectiveness of algorithms responsible for trajectory planing is largely dependent on the correct sensory input. An autonomous robotic platform must be equipped with distance sensors to allow object avoidance. With more complex environments like any outdoor terrain, standard point distance sensors are insufficient at high speeds. The solution lies with 2D and 3D laser range finders. An example of such a device is the Hokuyo UTM-30LX. With it's compact size it fits well in a mobile robotic platform, presented of Fig. 1.

Fig. 1 Autonomous robotic platform with Hokuyo UTM-30LX installed

To use any laser range finder, like UTM-30LX, in changing environments it is necessary to measure the influence of external conditions on the result. In the research done, Hokuyo laser range finder has been tested for: the drift effect, influence of target distance, surface brightness, color and material, and the sensor orientation.

The following measurements has been done to test the validity of using Hokuyo UTM-30LX on a mobile, wheeled robotic platform shown in Fig. 1.

2 Characterization of Laser Range Finders

The need of more accurate characterization of laser range finders was expressed in many works, including Desai et al [1], which expressed the need of proper selection of the device for best quality of measurements in performed task, or Pascoal et al [2], in which range finders were evaluated for their usefulness in specific mission. The specifications provided by the sensor's producer defines only general information about the device's accuracy and does not include details, such as the impact of different materials and colors on the measurement surface, drift effect, or the accuracy on different distances.

As autonomous vehicles are designed to operate in variety of environments, these characteristics need to enable the designer to predict possible inaccuracy of measurement and minimize it's impact on the vehicle's operation. In the most extreme situations, the device might not be able to detect certain materials. Pascoal et al in [2] showed, that Hokuyo URG-04LX was unable to detect black velvet, while Sick DT60 and IFM O1D100 results had a large error when detecting a reflective surface, such as a mirror.

Characterization of the device can also be used to create its model, either for calibration or simulation purposes. Propositions of such models were presented for other range finders: for Sick LMS 200 by Ye and Borenstein [3], and for Hokuyo URG-04LX by Kneip et al [4] and for UBG-04LX-F01 by Park et al [5].

3 Hokuyo UTM-30LX

Hokuyo UTM-30LX provides scanning range of 270° and 30 meters with angular resolution of 0.25° per step. It is able to measure distances up to 60 meters, but without guaranteed reliability. One full scan cycle lasts for 25 ms, which supplies a 40 Hz measurement frequency. Data transfer to host is realized through USB 2.0 interface, using dedicated SCIP 2.0 protocol in communication with device. Therefore it is necessary for robotic platform to implement such interface. It has been done using FTDI VNC1L 'Vinculum' USB Host Controller, which bridges the signal to STM32F microcontroller. This allowed the implementation of Ethernet (with provided Ethernet PHY) or CAN based communication onboard the mobile robot.

The measurement accuracy, according to the producer [11], is defined as $\pm 30\ mm$ at 0.1 to 10 m range, and $\pm 50\ mm$ at 10 to 30 m range. Precision of the repeated measurement, defined as a standard deviation of samples, is defined as less than 10 mm at 0.1 to 10 m range, and less than 30 mm at 10 to 30 m range. All those values apply to the measurement of white sheet.

Fig. 2 The setup of UTM-30LX device connected with STM32F based board with Ethernet connection to development computer

4 Results

All of the experiments were performed using the measurement on the front step of UTM-30LX, i.e. measurement of the single point located in front of the device.

For reception, processing and presenting data on host, the LabVIEW virtual instruments were used. Hokuyo provides the UTM-30LX driver for Windows and driver for their scanning laser devices for LabVIEW, therefore quick development of the testing applications was possible.

Nominal distance was measured, using Bosh DLE 70 Laser Rangefinder. It has typical measurement accuracy of ± 1.5 *mm*, which is one order of magnitude better than UTM-30LX, therefore it gives good reference distance value.

In each of the experiments, at least 1000 samples were taken. Result values formed a single, Gaussian-like lobe on the histograms.

The following subsections show the results of UTM-30LX characterization.

4.1 *Drift Effect*

To determine the measurement drift over time, the scanner was placed at about 2 meters distance from solid wall and set to make single measurement every half second. Before that, the device was powered off for at least 4 hours. The experiment was run for over 100 minutes in stable environment (i.e. ambient temperature and lighting). There was no need of more frequent sampling or more accurate distance referencing, as the drift was expected to be observed over longer than few minutes period and it should have approximately the same effect, regardless of the measured distance.

As shown on Fig. 3, during first 45 – 50 *min* of operation the measurement is rising about 8 – 10 *mm*. Two more repetitions of this tests confirmed the result.

To exclude drift effect from other tests, the device was running for at least 1 hour before each experiment.

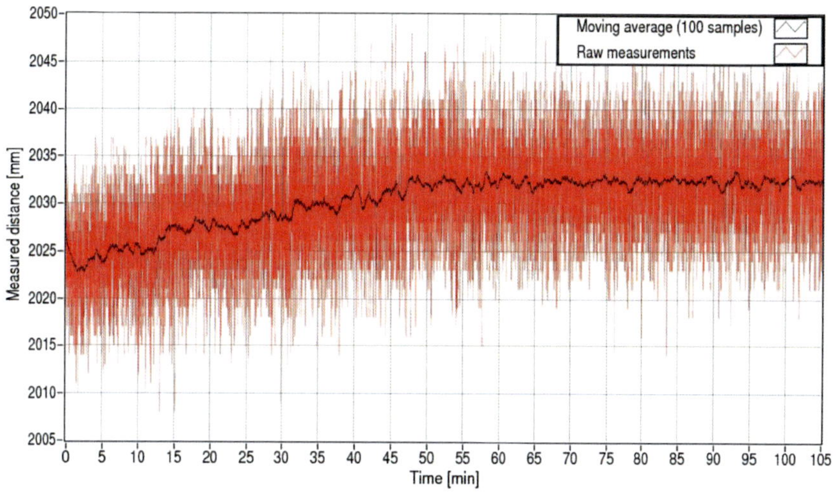

Fig. 3 Observed drift effect of measurements over time. Samples values are rising during the first 45 – 50 *min* of measurements.

4.2 *Target Distance*

Work of Kneip et al. [4] showed connection between measurement accuracy and distance of scanning device from the target. Park et al. [5] proposed also the investigation of relationship between standard deviation and distance. To determine those dependencies, 1000 samples were taken at every 2 meters distance, beginning at 2 *m*, up to 22 *m*, then mean value and standard deviation of measurements at each distance was calculated.

Standard deviation of measurements is rising with distance, as shown on Fig. 4. This relationship can be approximated with linear function. The most probable source of this effect is the increasing dissipation of laser light, resulting in lower intensity of captured reflection at higher distances. Such relationship was not so clearly observable in Parks et al. work, because of much shorter range of their measurements (only 91 – 4992 *mm*).

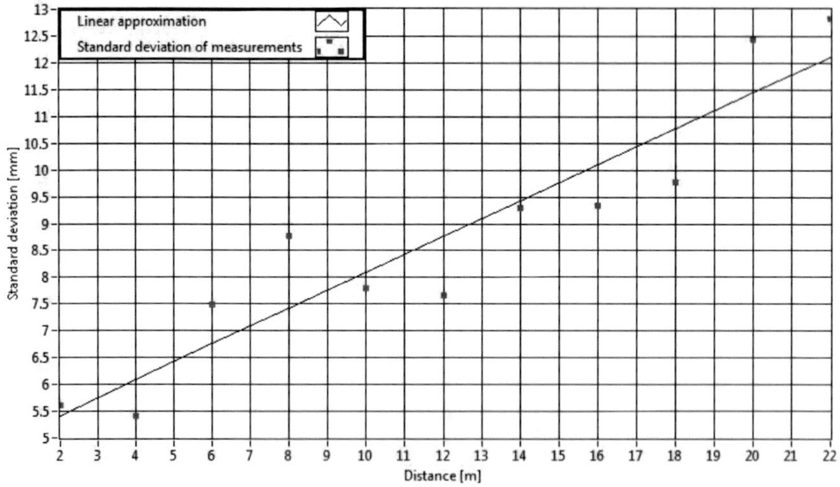

Fig. 4 Standard deviation of measurements over rising distance from target. Linear approximation was made using least squares method.

Absolute error, shown in Fig. 5, is not connected with the distance in any clear relationship, despite indication in device specification that accuracy at distance higher than 10 *m* should become worse. It is oscillating between −18 and 18 *mm*. As result, relative error is starting from 0.7% for 2 *m* and tending towards nearly 0% with increasing nominal distance.

4.3 *Target Surface Brightness, Color and Material*

The influence of target surface brightness and color was tested by placing the sheets of different brightness and colors at 1.5 meters distance from UTM-30LX. 5000 samples were taken with each brightness and color. Histograms of each series are shown in Fig. 6 and Fig. 7.

On the contrary to results of similar experiments performed with Hokuyo URG-04LX [4] and Hokuyo UBG-04LX-F01 [5], no clear relationship between surface

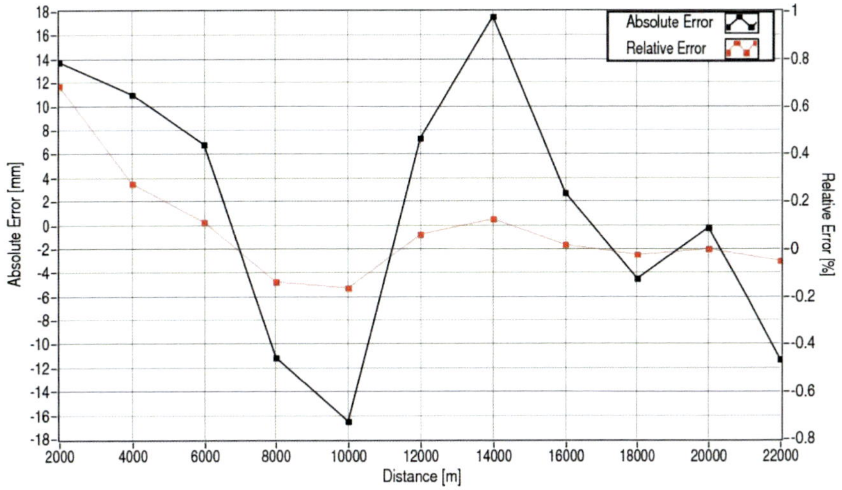

Fig. 5 Absolute error of measurements over rising distance from target

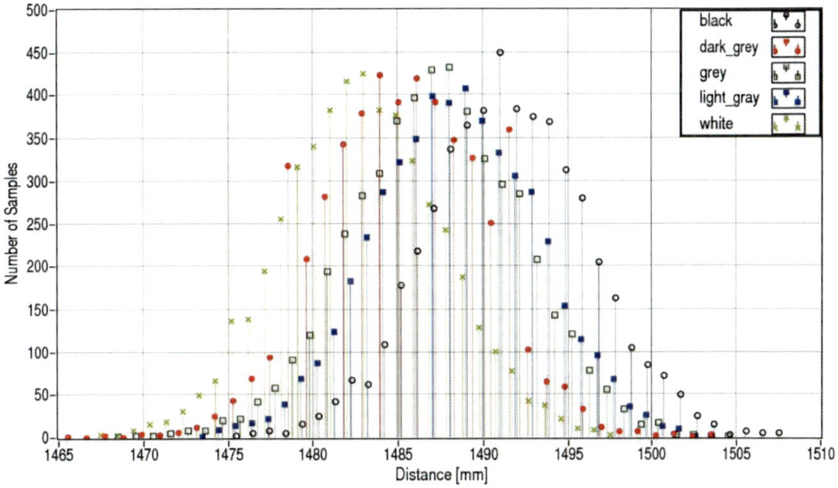

Fig. 6 Effect of brightness of target

brightness or color and samples values were showed for UTM-30LX. While the placement of histogram lobes for different brightness or colors can be distinguished for other devices, those lobes for UTM-30LX are nearly laying on each other. One possible explanation might be nearly six times higher range of UTM-30LX, which would mean higher energy of emitted laser light, so captured reflection should have better intensity, as the effect of scattering light on target surface at lower distances should be relatively less significant.

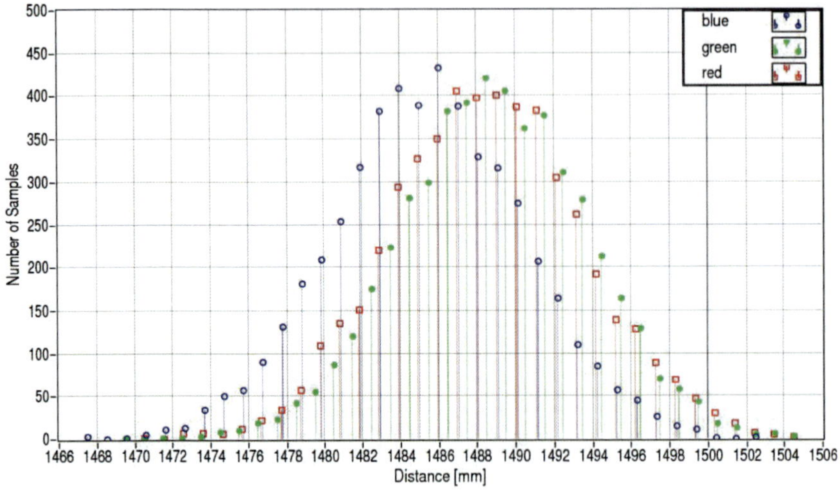

Fig. 7 Effect of color of target

Standard deviation and errors of measurements are shown in Table 1. Highest relative error was noted for the brightest (white) sheet: -1.14%, and lowest for black sheet: -0.56%. As for the surface color, green and red sheet was measured with same relative error of -0.77%, and blue sheet with slightly higher error of -0.97%. Most notably, all errors for color and brightness were negative.

The device showed very stable precision at this distance, disregarding target color and brightness, as standard deviation for each brightness and color is similar, oscillating around $4.80mm$.

Table 1 Summary of the measurements of different surface brightness and colors

	Absolute error [mm]	Relative Error [%]	Standard Deviation [mm]
White	-17.14	-1.14	4.83
Light Grey	-11.619	-0.77	4.87
Grey	-12.75	-0.85	4.80
Dark Grey	-14.78	-0.98	4.74
Black	-8.48	-0.56	4.87
Blue	-14.61	-0.97	4.86
Green	-11.59	-0.77	4.82
Red	-11.56	-0.77	4.87

Similar experiments were performed with different target materials. Chosen materials were: aluminium plate, steel plate with rougher surface, compact disk as example of very reflective material, moderately reflective plastic plate (green color), cotton cloth and wooden plank. The key to choose such materials were to check the effect of materials which might be meet by the device during operation on autonomous robot (wood, metal, clothes), or specific materials (CD, plastic).

Results of the experiment (Table 2) shows that reflectiveness of the material has the biggest impact on measurement accuracy and precision. The measurement of compact disk came with relative error of 1.82% and standard deviation of 9.13 *mm*, which are much higher than in any other performed experiment.

Table 2 Summary of the measurements of different materials

	Absolute error [mm]	Relative Error [%]	Standard Deviation [mm]
Aluminium	-6.74	-0.44	4.75
Steel	7.70	0.38	5.03
CD	27.39	1.82	9.13
Cotton	12.75	0.63	4.73
Wood	4.24	0.21	4.76
Plastic	7.25	0.36	5.10

4.4 *Scanner Orientation*

During operation in natural environment, scanner might be put in various orientations when robot is moving on rough terrain. This may influence the measurement, as gravity would have different impact on rotation of the sensor.

The experiment included measurements of target at 1.5 *m* distance from the device in four orientations: up (normal orientation), right (rotation of 90deg,

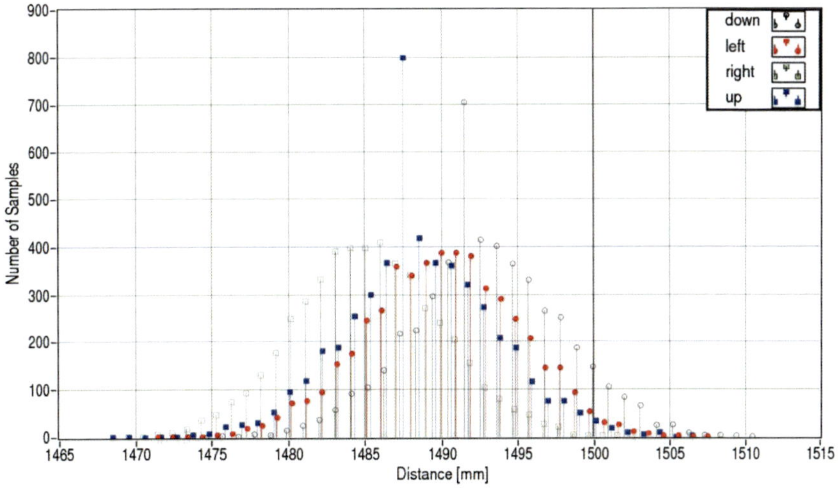

Fig. 8 Effect of orientation of the device

relative to normal orientation when looking from behind the device), down (rotation of 180deg) and left (rotation of 270deg). For each orientation 5000 samples were taken.

The most accurate measurements were made when the device was in 'down' orientation, with relative error of measurement of -0.47%, while the highest error was recorded in 'right' orientation: -0.99%. However, 'right' orientation also showed the least standard deviation of 4.95 *mm*, and the highest was for 'left': 5.11 *mm*. The differences in accuracy and precision for different orientations can be compared to those from the experiments with different colors and brightness.

5 Conclusions

The experiments performed and presented in this article have determined the usability of UTM-30LX laser rangefinder in autonomous mobile robots.

Results show that Hokuyo UTM-30LX can be used for distance scanning, targeting different materials, colors and brightness with similar effects. The relative errors of measurements were within $\pm 1\%$ and standard deviation was less than 5 *mm*. Such precision and accuracy is sufficient for usage in object avoidance algorithms. Caution must be taken only during measurements of highly reflective surfaces, because of the result higher error and standard deviation.

Because the sensor orientation doesn't have any significant impact on the results, Hokuyo UTM-30LX was positioned on the mobile robotic platform as shown in Fig. 1. It is necessary to take into consideration the influence of distance from the target, as the standard deviation of samples is rising linearly with the distance.

References

1. Desai, A., Huber, D.: Objective Evaluation of Scanning Ladar Configurations for Mobile Robots. IEEE/RSJ (2009), doi: 10.1109/IROS, 5354359
2. Pascoal, J., Marques, L., Almeida, A.T.: Assessment of laser range finders in risky environments. IEEE/RSJ (2008), doi: 10.1109/IROS, 4650961
3. Ye, C., Borenstein, J.: Characterization of a 2D laser scanner for mobile robot obstacle negotiation. IEEE (2002), doi: 10.1109/ROBOT, 1013609
4. Kneip, L., Tache, F., Caprari, G., Siegwart, R.: Characterization of the compact Hokuyo URG-04LX 2D laser range scanner. IEEE (2009), doi: 10.1109/ROBOT.2009.5152579
5. Park, C.-S., Kim, D., You, B.-J., Oh, S.-R.: Characterization of the Hokuyo UBG-04LX-F01 2D Laser Rangefinder. IEEE (2010), doi: 10.1109/ROMAN.2010.5598672
6. McDaniel, M.W., Nishihata, T., Brooks, C.A., Iagnemma, K.: Ground Plane Identification Using LIDAR in Forested Environments. IEEE (2010), doi: 10.1109/ROBOT.2010.5509963
7. Martinez, D.M., Haverinen J., Roning, J.: Sensor and connectivity board (SCB)for mobile robots. IEEE (2008), doi: 10.1109/BEC.2008.4657507
8. Surmann, H., Nuchter, A., Hertzberg, J.: An autonomous mobile robot with a 3D laser range finder for 3D exploration and digitalization of indoor environments. Elsevier (2003), doi: 10.1016/j.robot.2003.09.004

9. Lee, K.-H., Ehsani, R.: Comparison of two 2D laser scanners for sensing object distances, shapes, and surface patterns. Elsevier (2008), doi: 10.1016/j.compag.2007.08.007
10. Bellotto, N., Hu, H.: Multisensor-Based Human Detection and Tracking for Mobile Service Robots. IEEE (2009), doi: 10.1109/TSMCB.2008.2004050
11. Scanning range finder UTM-30LX product page. Hokuyo Automatic Co., Ltd., `http://www.hokuyo-aut.jp/02sensor/07scanner/utm_30lx.html` (cited July 13, 2011)

Direct Method of IMU Calibration

Witold Ilewicz and Aleksander Nawrat

Abstract. In the paper a method for fast IMU calibration is described. The method need previously calibrated IMU, called standard IMU. Standard IMU and IMU to be calibrated have to have a solid mechanical connection. Then a sequention of movement is done and an outputs from both IMUs are recorded simultaneously. Outputs can be recorded with different sampling frequency for each IMU. Then an iterative procedure is employed for simultaneous fitting signal from calibrated IMU to standard IMU and aligning both signals, using least squares method. The point in 2-parameter space, that give best fit allows to calculate a calibration matrixes that describes linear dependence beetwen calibrated and sandard sensors. A method allows to calibrate all three-axis sensors (accelerometers, angular rate sensors and magnetometers) at the same time. Method is not-time-consuming and can be used to convenient in-field and on-object calibration of IMUs without a complicated calibration setup.

Keywords: IMU, calibration, inertial, accelerometers gyroscopes, magnetic sensors, aligning, lsq minimization, model fitting.

1 Introduction

The market of inertial measurement units is constantly growing. Many even small institutions are able to design and complete project of IMU dedicated to specific application area. Usually IMU consist of 3-axis accelerometer, 3-axis gyroscope and 3-axis magnetometer. Additionally it can be equipped with barometric pressure

---

Witold Ilewicz · Aleksander Nawrat
Silesian University of Technology, Institute of Automatic Control,
Akademicka 16, 44-101 Gliwice, Poland
e-mail: {Witold.Ilewicz,Aleksander.Nawrat}@polsl.pl

A. Nawrat et al. (Eds.): Advanced Technologies for Intelligent Systems, SCI 440, pp. 155–171.
springerlink.com

sensor unit, temperature sensor unit and others sensors to ensure versatility and self-ability to errors compensation. After producing such IMU modules have raw outputs from A/D converters and need calibration. As a result many work is done to improve calibration methods dedicated to ineritial measurement units.

Calibration methods for inertial sensors generally need to compare output from sensor with known values generated by calibration setups and instruments. A different method were applied to ensure known quantity need to calibration. Optical trackers were used for example to calibration of IMU [8]. There were also used stands with precise maneuvers, with or without the Kalman filter use [3, 9, 10] to calibration of low-end inertial sensors for less demanding applications. Generally methods presented in literature has several drawback for users of low-cost IMUs. The main drawback is the necessity of the generation of accurate calibration values of acceleration, angular rate and so on. Other case is precision and recurrent mounting the IMU on the calibration stand and precise moving it during calibration. These methods usually requires expensive equipment to ensure required high-precision. For these reasons calibration procedure is usually expensive from point of view of small institutions. When done in third party laboratory, it is also time consuming. The precision at high level isn't always required. So there is a need to develop easy, fast and low-cost method for IMU calibration free of need of complicated setup. Attempts into this aim were undertaken in the works [11, 4, 5]. In the last one a method of calibration of accelerometers and gyroscopes only is presented without need of use of complicated setup. Method of calibration proposed in this article is consitent with the trends described above. It need only pre-calibrated IMU module. This module is then treated as a standard for uncalibrated IMUs. In the article a procedure is shown to calculate coeficients of calibration model of IMU on the base of standard IMU.

2 Calibration Procedure Description

Calibration procedure described in the article is understood as a set of steps leading to values of parameters of mathematical model that describes way of transforming raw digital outputs from sensors of uncalibrated IMU into value of measured quantity expressed in appropriate unit of measure.

The process of assigning the measure of x to the quantity y (where y is an output from uncalibrated system) is called the calibration of an measurement system [2]. Calibration may be performed by the indirect or direct method. The calibration is direct when the standard quantity aplied to the system to be calibrated is the same kind as the measured quantity. The characteristic of conversion is determined from the readings of output. Two possible cases of direct calibration are presented in Fig. 1. Raw output from modern uncalibrated sensor or transducer is usually an integer with range $\langle 0..2^R - 1 \rangle$ depending on resolution R of used A/D converter. Raw output from A/D converter has no phisical meaning before calibration and one can say

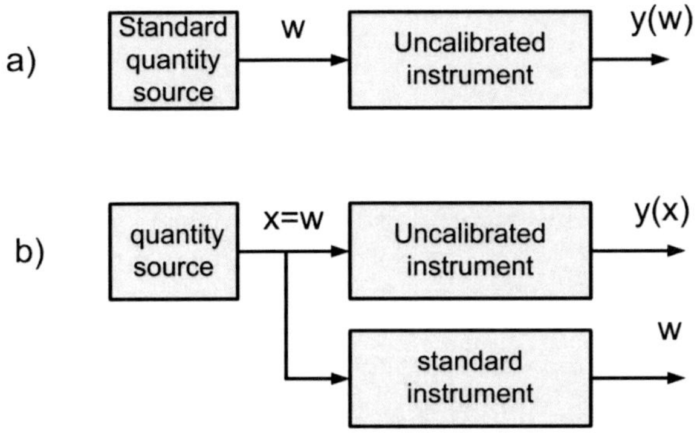

Fig. 1 Direct calibration. a) with a standard applied, b) with a standard instrument applied.

only that it is in some way proportional to measured quantity as acceleration, rate of rotation or magnetic field magnitude. Procedure of converting raw A/D converter output to measured quantity value is described by suitable mathematical model. Parameters of this model needs to be estimated. Method of calibration proposed in this work assumes, that there is possibility to collect simultaneously data from to be calibrated IMU and reference IMU (calibrated earlier). During data collection both IMUs are rigidly connected and stimulated by random or programmed movements, so data from both IMUs describe the same phenomenon. If a mathematical model describing dependence beetween data from uncalibrated IMU and standard (reference) IMU is known, data recorded in the way described just above can be used to estimate parameters of this model. Then model with known parameters can be used to calibrate data from uncalibrated IMU. Simplified diagram of calibration procedure described above is shown in Fig. 2.

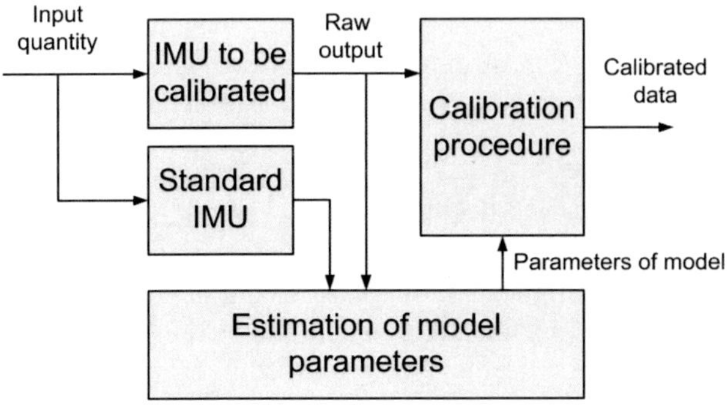

Fig. 2 Scheme of procedure of direct calibration of IMU sensors. Procedure is the same in case of accelerometers, gyroscopes and magnetometers.

2.1 Mathematical Model of Calibration

In direct calibration method, where 3-axis IMU is calibrated and 3-axis IMU is used as a standard (reference sensor) a linear model with 12 parameters can be used [1]. The model is the same for 3-axis accelerometers, gyroscopes and magnetometers.

Model that describes relationship beetwen output of standard IMU and output of uncalibrated IMU at a given moment of time for accelerometers can be expressed as

$$\begin{bmatrix} A_{Xcal} \\ A_{Ycal} \\ A_{Zcal} \end{bmatrix} = [MM]_{3x3} \cdot \begin{bmatrix} S_X & 0 & 0 \\ 0 & S_Y & 0 \\ 0 & 0 & S_Z \end{bmatrix} \cdot \left[\begin{bmatrix} A_X \\ A_Y \\ A_Z \end{bmatrix} - \begin{bmatrix} O_X \\ O_Y \\ O_Z \end{bmatrix} \right] \tag{1}$$

In equation (1) vector $\left[A_{Xcal}\ A_{Ycal}\ A_{Zcal} \right]^T$ is the output of 3-axis accelerometer in standard IMU and vector $\left[A_X\ A_Y\ A_Z \right]^T$ is the output of 3-axis accelerometer in uncalibrated IMU, vector $\left[O_X\ O_Y\ O_Z \right]^T$ is a set of offsets for 3-axis accelerometer in uncalibrated IMU, matrix $[MM]_{3x3}$ is misaligment rotation matrix between axes of 3-axis accelerometers in both IMUs and diagonal matrix with elements S_X, S_Y, S_Z is a matrix of gains for 3-axis accelerometer in uncalibrated IMU. As mentioned above model described by equation (1) is valid for measurements from both IMUs taken at the same time moment. During calibration a number of N measurements is taken. Each measurement taken at time moment t_i where $i = 1..N$ can be presented in form of 3-D row vector for both IMUs. So measurement for both IMUs taken during calibration expreriment can be set in form of $Nx3$ matrixes, in which each row is a single measurement taken at time moment t_i. If measurements for both IMUs are taken at the same moments of time, equation (1) can be rewriten as

$$\left[A_{Xcal}\ A_{Zcal}\ A_{Zcal} \right] = \left[A_X\ A_Y\ A_Z\ 1 \right] \cdot \begin{bmatrix} K_{11} & K_{12} & K_{13} \\ K_{21} & K_{22} & K_{23} \\ K_{31} & K_{32} & K_{33} \\ K_{41} & K_{42} & K_{43} \end{bmatrix} . \tag{2}$$

Left side of the equation (2) is matrix $Nx3$ of measurements from reference (standard) IMU. On the rigth side there is matrix $Nx4$ where first three columns are measurements from uncalibrated IMU and fourth column has all elements equal to 1. Matrix with elements K_{ij} (where $i = 1..4$ and $j = 1..3$) is a calibration coefficients martix. Equation (2) is description of calibration procedure. In other words, if one know values of elements $[K_{ij}]$ of the matrix on the right side of equation (2), calibration requires just right-hand multiplication of measurement matrix from uncalibrated IMU with added column of 1 by a matrix of calibration coefficients $[K_{ij}]$. So matrix $[K_{ij}]$ is a set of calibration coeficients to be estimated in calibration procedure. Coeficients in martix $[K_{ij}]$ are directly related to coeficients in equation (1). Equation (2) is linear and can be expressed in simplier form

$$[A_{CAL}]_{Nx3} = [A]_{Nx4} \cdot [K]_{4x3}, \tag{3}$$

where A_{CAL} is $Nx3$ matrix of measurements from standard IMU, A is $Nx3$ matrix of measurements from uncalibrated IMU with added column of "1" and K is $4x3$ matrix of unknown calibration coefficients. The K matrix can be estimated in one step using LSQ method and the result is:

$$K = \left[A^T A\right]^{-1} A^T A_{CAL}. \tag{4}$$

So simple solution of problem of estimation of calibration coefficients can be used only in case that measurements (rows) in matrixes A_{CAL} and A are taken at the same time moment. In practice this is very difficult to ensure. Usually data from both IMUs are recorded independly at different rates and synchronization of clocks in IMUs is practically impossible. As a result matrices A_{CAL} and A have different length (different number of rows) and direct use of equation 4 is impossible. synchronization of signals recording from both IMUs could be a problem too. Usually a special setup is needed to ensure the same start and stop time of signal recording for both IMUs during excitation experiment. If IMUs have different intefraces and are connected to PC computer via different USB ports (or similar interface), synchronization of records from both IMUs is practically impossible. So procedure of estimation of calibration coefficients should be extended to ensure synchronization of signals from both IMUs.

2.2 *Estimation of Calibration Coefficients with Autosynchronization of Signals*

During calibration experiment there are recorded two signals - one from standard (reference) IMU and second from noncalibrated IMU. Signals are discrete and sampled with different sampling frequency. Start times of both records are different (lack of synchronization) and length of time of record is different too. Of course records have different number of samples. Both recorded signals are description of the same phisical phenomena - changes of acceleration, magnetic field or angular rate in time. So shape of both records should be similar. This similarity should be used to synchronize both signals.

Fig. 3 is ilustration of sampling during calibration experiment. Input to both IMUs is the same. As it could be seen in Fig. 3a) standard and uncalibrated IMUs start sampling at different time moments (respectively t_{1s} and t_1) so $t_1 \neq t_{1s}$. Both IMUs finish sampling at different time moments too (respectively t_{Ns} and t_q) so $t_q \neq t_{Ns}$. Sampling intervals are different for IMUs. In order to compare the signals during calibration procedure, signal from uncalibrated IMU should be transformed to time scale of standard IMU. This transformation is done in the way that time of the first sample of signal from uncalibrated IMU is set equal to time of the first sample of signal from standard IMU. Similary, time of last sample of signal from uncalibrated IMU is set as equal to time of the last sample from standard IMU. Time

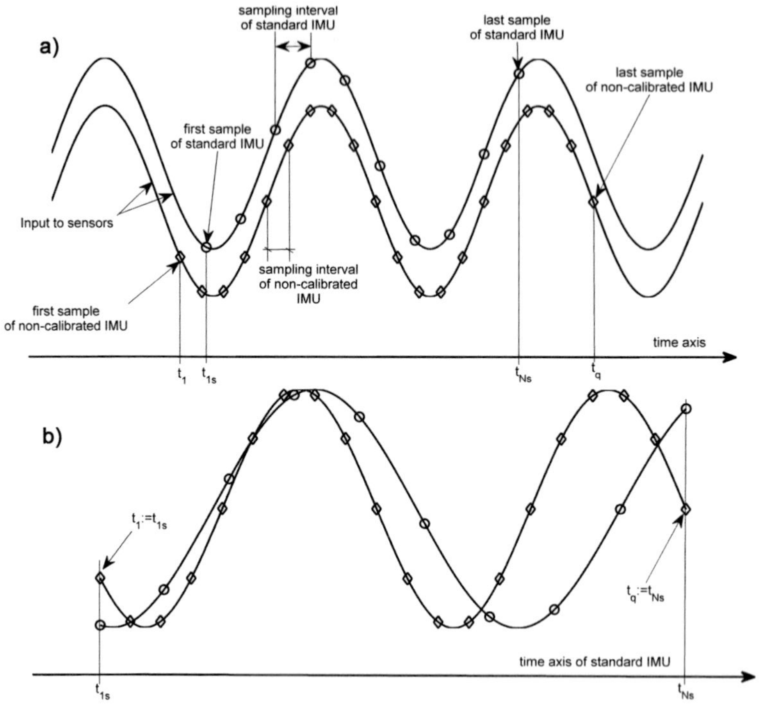

Fig. 3 Example of typical sampling results durnig calibration experiment for single 1-D sensor a) sampling procedure for standard (upper) and uncalibrated (lower) IMUs; b) comparison of output records from both IMUs after rescaling time axis of uncalibrated IMU record. For simplicity, both outputs are presented in the same scale of ordinates.

moments assigned to all samples in signal from uncalibrated IMU that lie between first and last sample are then rescaled assuming uniform sampling. After rescaling time axis of signal from uncalibrated IMU, both signals can be plot in time scale of standard IMU as shown in Fig. 3b). Record from standard IMU is basis of calibration. It consists of 10 samples and defines range of input used to calibration. Range of input included between first and last sample for uncalibrated IMU is different from range for standard IMU (because of lack of synchronization of sampling) and number of samples is 20 and different from number of samples for standard IMU because of different sampling frequency. In that case procedure of estimation of calibration matrix coefficients described by Eq. (4) is impossible to perform. Before use it, signal from uncalibrated IMU should be resampled in the way that after resampling is has the same number of samples the same as signal from standard IMU and should be cut to range of time the same as for signal from standard IMU. In other words it should be synchronized with signal from standard IMU.

Resampling of signal from uncalibrated IMU

Resampling of signal from uncalibrated IMU could be done by use of cubic spline interpolation. Signal recorded from uncalibrated IMU is then replaced by values of spline in points equally distributed between first and last sample of signal from uncalibrated IMU and number of points of spline is equal to number of points in signal from standard IMU. After that number of samples from both IMUs is the same - so formally formula (4) can be used. Example of resampling for signals from Fig. 3b) is presented on Fig. 4.

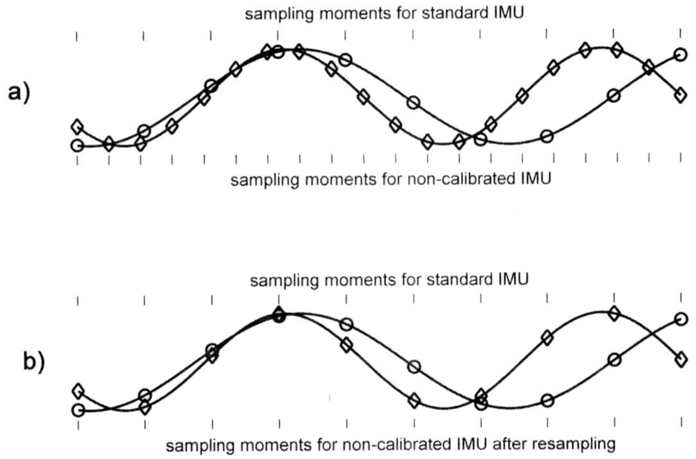

Fig. 4 Example of resampling procedure a) comparison of records before resampling b) comparison of records after after resampling

After resampling signal from uncalibrated IMU Eq. 4 can be formally used to calculate values of elements of calibration matrix K_{ij}. Of course result of estimation of K_{ij} is optimal in case when first and last points of signal resampled from uncalibrated IMU are taken in the same time point as first and last sample of signal from standard IMU, in other words a conditions $t_l = t_{1s}$ and $t_q = t_{Ns}$ (Fig. 3a) must be fulfilled.

Synchronization of signal from uncalibrated IMU to signal from standard IMU

As a measure of quality of synchronization of both signals can be used sum of squares of differences between signal from standard IMU and signal from uncalibrated IMU transformed using Eq. 2 by actual matrix of calibration coefficients calculated according to Eq. 4. This quality indicator has minimal value in case that points of standard signal and signal from uncalibrated IMU after resampling and calibration are ideally synchronized.

Let A_S denotes Nx3 matrix of measurements from 3-D sensor (accelerometer, magnetometer or gyro) of standard IMU, where each row is a single measurement and first row it taken at time t_{1s}. Last row is taken at time t_{Ns} and all rows are taken with sampling period T_s. Corresponding time scale is described by time vector $t_S = [t_{S1},..,t_{SN}]$ whose elements describes time moments of sampling rows of matrix A_S. Without loss of generality elements of vector t_S can be replaced with integers from 0 to $N-1$ and this normalized time vector $t_{0:N-1} = [0,...,N-1]^T$ is further treated as a new generalized time scale of signal from standard IMU.

Let A denotes qx3 matrix of outputs from 3-D corresponding sensor of uncalibrated IMU, where each row is a single output and first row it taken at time t_1. Last row is taken at time t_q and all rows are taken with sampling period T (in general $T \neq T_s$) so time vector corresponding to A matrix is $t = [t_1,..,t_q]^T$. Assume that $t_1 < t_{1s}$ and $t_q > t_{Ns}$ so that time domain of signal recorded from standard IMU is included in time domain of output from uncalibrated IMU (like in example on Fig. 2a). In order to perform calibration matrix A should by coverted into martix A^R with size the same as size of matrix A_S in the way descripted below.

Let intruduce two decision variables t_b and t_e. These variables have the following properties: $t_1 \leq t_b < t_e \leq t_q$. Variables t_b and t_e will be used to ensure synchronization of signals from standard IMU and from uncalibrated IMU. The interpretation of these variables is shown on Fig. 5. Cubic spline built on the time domain of recorded output from uncalibrated IMU allows to calculate value of the output at any time point between t_1 and t_q. Points t_b and t_e in the time domain of uncalibrated IMU signal are aproximation of points t_{1s} and t_{Ns} in time domain of standard IMU signal respectively. For fixed values of t_b and t_e matrix A^R is constructed from matrix A in the way that first row of matrix A^R are values of cubic spline calculated at t_b and the last row of matrix A^R are values of cubic spline calculated at t_e and all other elements of matrix A^R are calculated as a values of spline at $N-2$ time points equally spaced between t_b and t_e, so number of rows of matrix A^R is equal to number of rows of matrix A_S. Now to each row with number i of matrix A^R can be assigned value $i-1$ of time vector $t_{0:N-1} = [0,...,N-1]^T$ assigned to matrix A_S. Matrices A_S and $A^R = A^R(t_b,t_e)$ are now outputs of standard 3-D sensor and uncalibrated 3-D sensor respectively and relationship between this two matrices is described by Eq. 2. Using result from Eq. 4 output from uncalibrated IMU can be converted into calibrated data A_C^R according to

$$A_C^R = A_1^R\left[\left(A_1^R\right)^T A_1^R\right]^{-1}\left(A_1^R\right)^T A_S, \tag{5}$$

where matrix A_1^R is composed from matrix A^R of size Nx3 and vector of 1 of size Nx1:

$$A_1^R = A_1^R(t_b,t_e) = [A^R(t_b,t_e)1]. \tag{6}$$

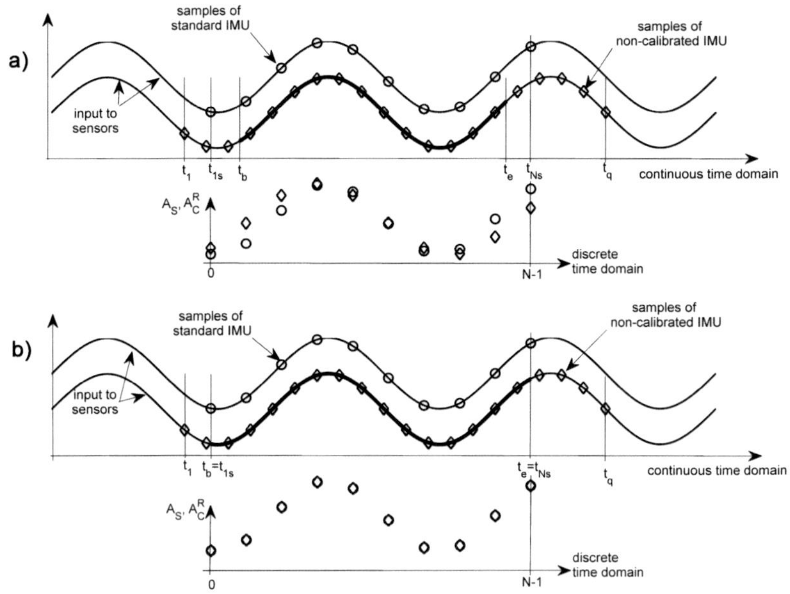

Fig. 5 Interpretation of t_b and t_e variables. a) not optimal choosing of t_b and t_e; b) optimal choosing of t_b and t_e.

Matrix A_C^R has size Nx3 and can be directly compared with matrix A_S so quality indicator of synchronization of signals recorded in both matrix can be expressed as

$$J(t_b,t_e) = \sum_{j=1}^{3}\sum_{i=0}^{N-1}\left[[A_S]_{ij} - \left[A_C^R(t_b,t_e)\right]_{ij}\right]^2. \tag{7}$$

Synchronization procedure can be performed by minimizing Eq. 7 according to variables t_b and t_e:

$$\hat{t}_b,\hat{t}_e = \arg\min_{t_b,t_e} J(t_b,t_e) \tag{8}$$

Values $\hat{t}_b$ and $\hat{t}_e$ that minimizes J ensure the synchronization of both signals, so time range of output from uncalibrated IMU used during calibration is the same as time range of output from standard IMU. Calibration matrix $\hat{K}$ that converts outputs from uncalibrated IMU into calibrated data can be determined immediately as:

$$\hat{K} = \left[\left[A_1^R(\hat{t}_b,\hat{t}_e)\right]^T A_1^R(\hat{t}_b,\hat{t}_e)\right]^{-1}\left[A_1^R(\hat{t}_b,\hat{t}_e)\right]^T A_S, \tag{9}$$

where $\hat{K}$ is 4x3 calibration coefficient matrix. Finally calibration procedure that converts raw single output from 3-D sensor from uncalibrated IMU $\left[A_X\ A_Y\ A_Z\right]^T$ taken at time t_0 into calibrated measurement with unit is described by:

$$\begin{bmatrix} A_X^{\hat{K}} \\ A_Y^{\hat{K}} \\ A_Z^{\hat{K}} \end{bmatrix} = \hat{K}^T \cdot \begin{bmatrix} A_X \\ A_Y \\ A_Z \\ 1 \end{bmatrix}. \quad (10)$$

3 Results and Discussion

Verification of the algorithm described in the preceding paragraph was based for applying this algorithm to calibrate prototype IMU module from the test series of modules designed to use on wheeled unmanned vehicle. This prototype IMU module will be called in the following section as CZK. CZK contains a 3-axis acceleration sensor, 3-axis magnetic sensor and 3-axis angular rate sensor. Output of the CZK is generated with rate aproximately 200Hz and before calibration output from each single sensor of the IMU is a number from range $< 0..2^{16} - 1 >$ because of use 16-bit A/D converters. As a standard IMU MTiG-28 manufactured by Xsens was used with corresponding sensors onboard. MTiG-28 IMU module will be called in the following section as MTI. Both IMUs and their calibration configuration are presented on Fig. 6. During calibration IMUs were rigidly connected via aluminium plate with the use of glue. Calibration experiment lasted aproximately 30s. In the

Fig. 6 Calibration configuration of IMUs during calibration experiment. As a standard MTiG-28 was used (orange).

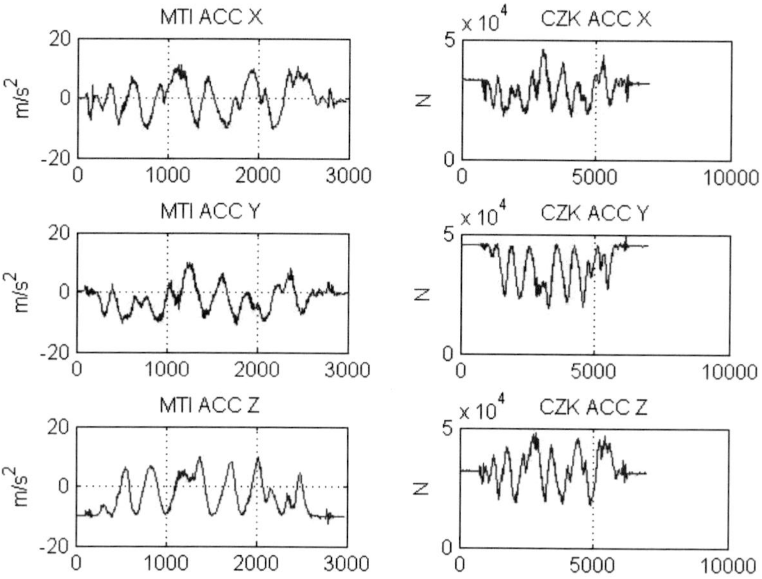

Fig. 7 Data recorded during calibration experiment for accelerometers. On the left side data from MTI accelerometers, on the right side data from CZK accelerometers.

experiment both IMUs were were stimulated with sequence of random movements. During experiment data from both IMUs were recorded. Standard IMU data were sampled with nominal frequency equal to 100Hz and output was written to file on the mass storage. Data from uncalibrated IMU were sampled with unknown frequency about 200Hz. Both records included data from 3-axis accelerometer, 3-axis magnetometer and 3-axis gyroscope as the aim of calibration was to calibrate all sensors in CZK. There has been no action to ensure the synchronization of records during experiment. Data recorded during experiment are presented from Fig. 7 to Fig. 8 for 3-axis accelerometers, 3-axis gyroscopes and 3-axis magnetometers respectively. During calibration experiment about 3000 samples from standard IMU were recorded. Because of higher sampe frequency, number of samples recorded from uncalibrated IMU was about 6000. In the algorithm of calibration it is required to determine values of t_{1s} and t_{Ns} for the standard IMU and values of t_1 and t_q for record of output from uncalibrated IMU. In case shown on Fig. 7 - Fig. 11 t_1 expressed as a number of sample was set to value 247 and t_q was set to 2588. Parameters of output record from uncalibrated IMU were set $t_1 = t_b = 1044$ and $t_q = t_e = 5716$.

Iterative Nedler-Mead simplex minimization algorithm was used to minimize indicator of quality J described by equation 7. Initial values of parameters t_b and t_e

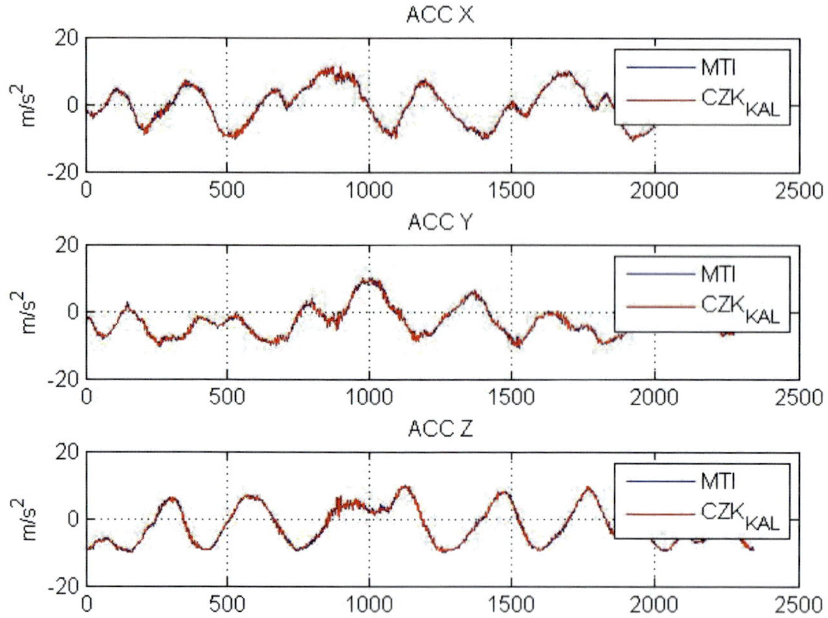

Fig. 8 Result of calibration of accelerometers. Comparison of data from standard MTI IMU and data from CZK after calibration. CZK_{KAL} are data after calibration.

were set as just above. As a result of minimization of J optimal values $\hat{t}_b = 1041.643$ and $\hat{t}_e = 5718.044$ were obtained. Below there are presented results of calibration for accelerometers, gyroscopes and magnetic sensors.

Results of accelerometeres calibration

Data used to 3-axis accelerometer calibration are shown on Fig. 7. It can be seen that there are corelation between axis of standard and uncalibrated IMU acceleration sensors - they are aproximately parallel in pairs: MTI X || CZK Z with the same direction of sensors; MTI Y || CZK X with the same direction of sensors; MTI Z || CZK Y with opposite direction of sensors. Generally the relative orientation of sensors in both IMUs doesn't matter for calibration results but if axes of sensors in IMUs are approximately parallel, interpretation of results is simplier and square matrix built of three first rows of calibration matrix is closer to diagonal matrix. Values of elements of calibration matrix for accelerometers obtained during the calibration are shown below:

$$K_{ACC} = \begin{bmatrix} 0.0116 \cdot 10^{-3} & 0.7446 \cdot 10^{-3} & -0.0115 \cdot 10^{-3} \\ 0.0201 \cdot 10^{-3} & -0.0069 \cdot 10^{-3} & -0.7537 \cdot 10^{-3} \\ 0.7471 \cdot 10^{-3} & -0.0098 \cdot 10^{-3} & 0.0182 \cdot 10^{-3} \\ -25.5022 & -23.4992 & 24.1625 \end{bmatrix} \quad (11)$$

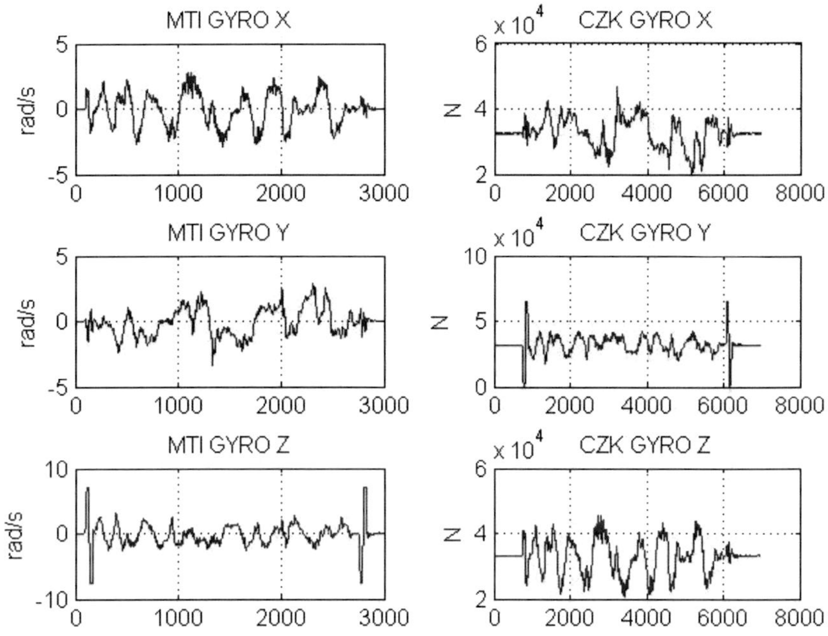

Fig. 9 Data recorded during calibration experiment for gyroscopes. On the left side data from MTI gyroscopes, on the right side data from CZK gyroscopes.

Results of use calibration matrix from equation 11 to convert data from uncalibrated IMU are shown in Fig. 8.

Results of gyroscopes calibration

Data used to 3-axis angular rate sensor (gyroscope) calibration are shown on Fig. 9. It can be seen that there are corelation between axis of standard and uncalibrated IMU acceleration sensors -as for accelerometers in previous paragraph - they are aproximately parallel in pairs: MTI X || CZK Z with the same direction of sensors; MTI Y || CZK X with opposite direction of sensors; MTI Z || CZK Y with opposite direction of sensors. Values of elements of calibration matrix for gyroscopes obtained during the calibration are shown below:

$$K_{GYRO} = \begin{bmatrix} -0.0029 \cdot 10^{-3} & -0.2304 \cdot 10^{-3} & -0.0003 \cdot 10^{-3} \\ 0.0054 \cdot 10^{-3} & -0.0000 \cdot 10^{-3} & -0.2321 \cdot 10^{-3} \\ 0.2263 \cdot 10^{-3} & -0.0056 \cdot 10^{-3} & 0.0006 \cdot 10^{-3} \\ -7.5757 & 7.6726 & 7.3920 \end{bmatrix} \quad (12)$$

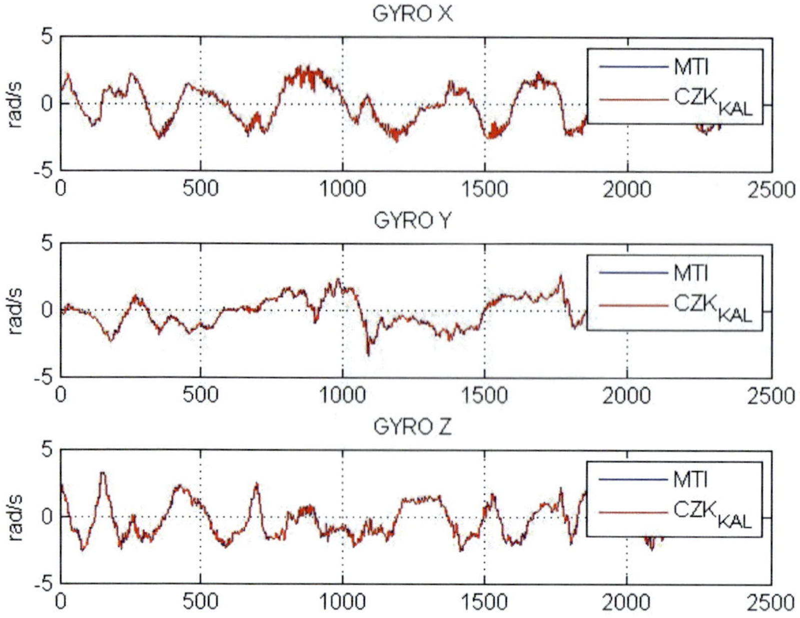

Fig. 10 Result of calibration of angular rate sensor. Comparison of data from standard MTI IMU and data from CZK after calibration. CZK$_{KAL}$ are data after calibration.

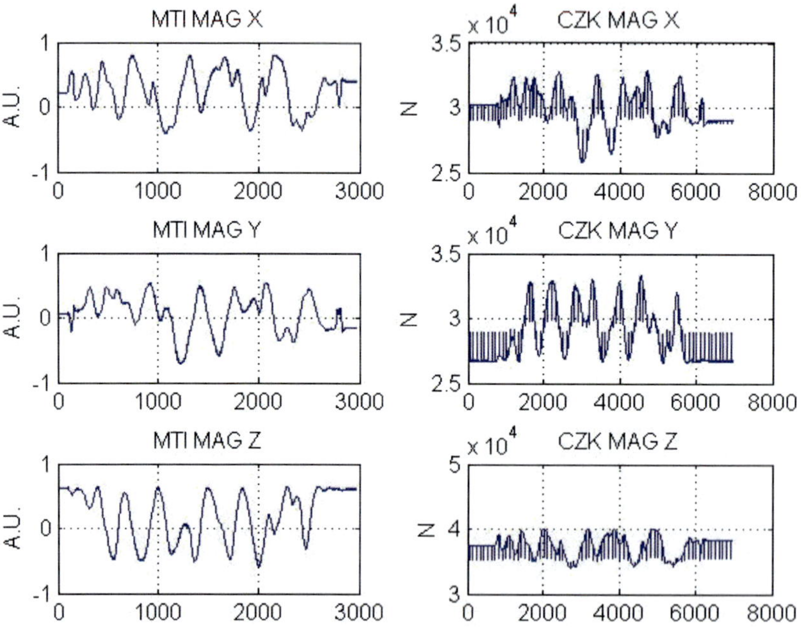

Fig. 11 Data recorded during calibration experiment for magnetic sensors. On the left side data from MTI magnetic sensors, on the right side data from magnetic sensors of CZK.

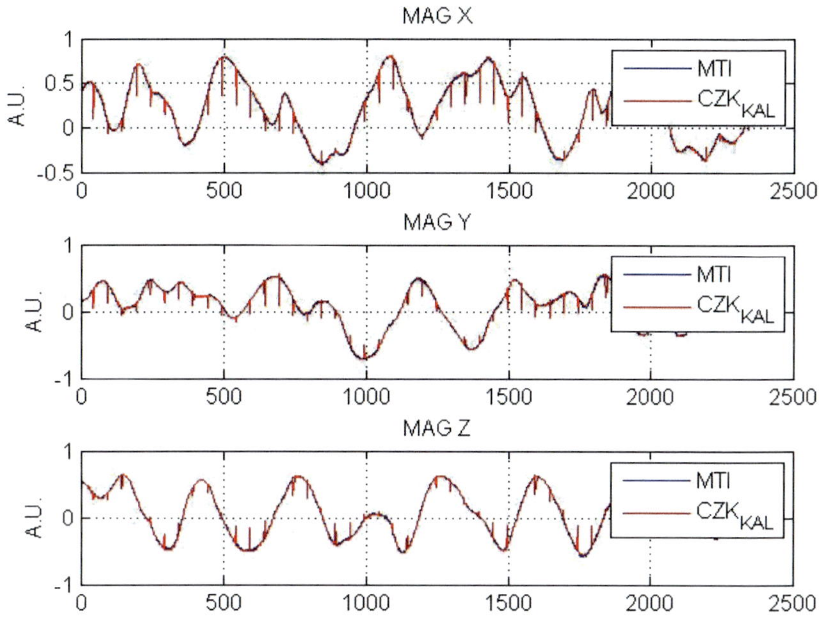

Fig. 12 Result of calibration of 3-axis magnetic field sensor. Comparison of data from standard MTI IMU and data from CZK after calibration. CZK_{KAL} are data after calibration.

Results of use calibration matrix from equation 12 to convert 3-axis gyroscope data from uncalibrated IMU are shown in Fig. 10. After calibration both signals visible on the figure are nearly identical.

Results of magnetic sensors calibration

Data used to 3-axis magnetic field sensor (magnetometer) calibration are shown on Fig. 11. There are visible spikes of noise on the output from CZK - spikes of noise are present because of repeated in time procedure of degaussing of magnetic sensors written to the firmware of CZK.

It can be seen that there are corelation between axis of standard and uncalibrated IMU acceleration sensors -as for accelerometers and magnetometers in previous paragraph - they are aproximately parallel in pairs: MTI X || CZK Z with the same direction of sensors; MTI Y || CZK X with the same direction of sensors; MTI Z || CZK Y with opposite direction of sensors.

Values of elements of calibration matrix for magnetic field sensors obtained during the calibration are shown below:

$$K_{MAG} = \begin{bmatrix} -0.0043 \cdot 10^{-3} & 0.1811 \cdot 10^{-3} & 0.0022 \cdot 10^{-3} \\ 0.0184 \cdot 10^{-3} & 0.0037 \cdot 10^{-3} & -0.1841 \cdot 10^{-3} \\ 0.1961 \cdot 10^{-3} & 0.0154 \cdot 10^{-3} & -0.0089 \cdot 10^{-3} \\ -7.4868 & -6.0724 & 5.8159 \end{bmatrix} \quad (13)$$

Results of use calibration matrix from equation 13 to convert 3-axis magnetic sensor data from uncalibrated IMU are shown in Fig. 12. After calibration both signals visible on the figure are nearly identical.

4 Conclusions

Presented in the article metod of IMU calibration has many advantages. First of all the method doesn't need complicated and expensive setup. The only thing necessary to use it is calibrated IMU and personal computer with proper to IMU interfaces.

Method can be complementary to other methods of IMU calibration that uses complicated setups. In that case metod can be used to confirm that accuracy of calibration setup is at the desired level or can be easily used to checking of calibration quality and to detect calibration faults.

Unlike many calibration methods, method presented in this article allows to simultaneous calibration of all types of sensors included in an IMU under condition that uncalibrated and standard IMUs have the same kinds of sensors onboard.

Many IMU calibration methods uses complicated setups to set IMU in known orientation with desired accuracy. There are known setups with robots used to determine desired position of IMU during calibration with desired accuracy. Usually the accuracy is ensured only in stationary points - when IMU doesn't move - because of unknown dynamics of robot ([3]). Application of standard IMU allows use of previously unusable measurements done during changing IMU position.

Other advantage of the method presented in this article is possibility to use method for purpose of simultaneous calibration of many sensors - and ensures easy way to check homogeneity of a series of sensors produced in single process.

Because method presented in the article doesn't need complicated setup, it can be used to fast calibration in field, where it is impossible use of dedicated to calibration setup.

References

1. Using LSM303DLH for a tilt compensated electronic compass, AN3192 Application note, Doc ID 17353 Rev. (August 1, 2010),`http://www.st.com`
2. Piotrowski, J.: Theory of physical and technical measurement. Elsevier Polish Scientific Publ. (1992)
3. Ilewicz, W., Nawrat, A.: Application of robot stand to the study of IMU modules accuracy - testing of angular rate sensors. In: Technological Advancement in State Defence and National Security. WPKJS, Gliwice (2010)

4. Skog, I., Handel, P.: Calibration of a MEMS inertial measurement units. In: XVII Imeco World Congress, Metrology for a Sustainable Developtment, Rio de Janeiro, Brasil (2006)
5. Fong, W.T., Ong, S.K., Nee, A.Y.C.: Methods for in-field user calibration of an inertial measurement unit without external equipment. Meas. Sci. Technol. (2008), doi: 10.1088/0957-0233/19/8/085202
6. Dorveaux, E., Visiere, D., Martin, A.-P., Petit, N.: Iterative calibration method for inertial and magnetic sensors. In: Joint 48th IEEE Conference on Decision and Control and 28th Chinese Controlo Conference, Shanghai, P.R. China (2009)
7. Hall, J.J., Williams II, R.L., van Graas, F.: Inertial Measurement Unit Calibration Platform. In: Sixth Conference on Applied Mechanisms and Robotics, Cincinnati, OH (1999)
8. Kim, A., Golnaraghi, M.: Initial calibration of an inertial measurement unit using an optical position tracking system. In: Proc. IEEE Plans, Monterey, April 26-29, pp. 96–101 (2004)
9. Grewal, M., Henderson, V., Miyasako, R.: Application of Kalman filtering to the calibration and alignment of inertial navigation systems. IEEE Trans. Autom. Control 36, 4–13 (1991)
10. Foxlin, E., Naimark, L.: Miniaturization, calibration and accuracy evaluation of a hybrid self-tracker. In: Proc. 2nd IEEE/ACM ISMAR, Tokyo, October 7-10, pp. 151–160 (2003)
11. Lotters, J.C., Schipper, J., Veltink, P.H., Olthuis, W., Bergveld, P.: Procedure for in-use calibration of triaxial accelerometers in medical application. Sensors and Actuators A: Physical 68(1-3), 221–228 (1998)

Part IV
Construction and Implementation of the Base Station and Software for Autonomous Objects

Nowadays, design and implementation of IT project consisting of both hardware construction and software implementation is a challenging task. This chapter will present some of the important issues in the field of Research and Development. It is known that the purpose of a commercial project is delivery of a ready product compliant with specific client's functional and non-functional requirements. Specific type of projects are R&D projects because each of such projects requires both scientific and technical knowledge in order to verify the hypothesis specified in the project's vision. Such projects are highly challenging, mostly due to client ambiguity, lack of indicated schedule and problems with defining partial aims. During implementation of the software presented in this chapter the best behavior patterns were adapted from known commercial projects. During work in uncertain environment agile design methodologies are assumed as the best ones. It is presented that application of such methodologies allows to build a highly motivated team whose goal has been to improve the quality of autonomous algorithms regardless of the environment conditions.

Another important topic presented in the chapter is configuration of infrastructure as well as hardware and software platform for the needs of the named base station system constituting a part of commanding unmanned objects. Nowadays, a presentation to end users layer is becoming more and more important. An example how to design such graphical user interfaces shall be presented on the example of steering and monitoring application of the command support system of unmanned objects. During the design phase of the project it was decided that an impact should be placed upon allowing the end user to manage the object in an optimal manner and minimize the time of delivery and display. The solution should be based upon one of the widely accessible software technologies (Java, Jboss, .NET) and that the system may be used to command various types of objects such as e.g. helicopters, ground vehicles, airplanes, flying wings and objects designed to operate in water.

Main Aims and Objectives of an IT System in the Implementation of the Project: Design and Implementation of Innovative Unmanned Mobile Platforms for the Needs of Monitoring State Borders

Marcin Błaszczok, Arkadiusz Gwóźdź, Łukasz Hoppe, and Krzysztof Skórka

Abstract. The subject of the project is to develop, design and implement a prototype of an unmanned aircraft with an IT system to support the management and monitoring of its parameters, whose main purpose is to monitor the borders of the state. The main goal of the project is to prepare such a computer system that would enable comprehensive management and monitoring of mobile devices.

Keywords: GIS Web application, setting routes and tasks, system GRANICE.

1 Introduction

The remainder of this paper presents the elements of the prototype system developed to enable comprehensive surveillance and monitoring of unmanned autonomous flying platforms. The main objectives of the IT system developed under the project are:

- Collection telemetry data from the flying object,
- Visualization of the telemetry data,
- Presentation and archiving of images from the platform camera,
- Steering the object in real time,
- Setting routes and tasks,
- Integration with other IT systems.

The main works that were carried out under the project together with the technology employed and goals will also be described. The main works under the project are:

- Analysis of the requirements,
- R & D works,

Marcin Błaszczok · Arkadiusz Gwóźdź · Łukasz Hoppe · Krzysztof Skórka
WASKO S.A., Berbeckiego 6, 44-100 Gliwice, Poland
e-mail: {m.blaszczok,a.gwozdz,l.hoppe,k.skorka}@wasko.pl

A. Nawrat et al. (Eds.): Advanced Technologies for Intelligent Systems, SCI 440, pp. 175–183.
springerlink.com

- System development,
- System tests.

The scope of the above work includes the issues related to the ESB, GIS, BI and reporting systems. It will also be necessary to develop appropriate communication protocols for processing the telemetry of the unmanned aircraft. A prototype system will be developed with the SOA technology, and divided into functional modules. The data from all modules will be collected in a central database that will enable data exchange with other systems through the selected integration bus. The main components include:

- Map Module,
- Control Module,
- Virtual Cockpit,
- Communication Module,
- Video module.

Besides the main elements of the system, also a dedicated application for mobile devices will be developed. This application will enable the tracking of object position on the map, the projection of images from cameras and issuing simple navigation commands.

2 Map Module

The software for a monitoring station will be designed to work with any number of unmanned objects working in the field. It will introduce significant improvements, of which one of the most important is an automated reporting of the discovery of the people in the border area, or the detection of suspicious objects. The possibility of monitoring from the superior unit level, will allow more effective detection and will minimize illegal border crossing or smuggling of illegal substances and materials. In accordance with the project, each institution of the Border Guard will be equipped with a PC running appropriate software, together with a map, which covers an area of responsibility within the scope of business of the unit.

The terrain map on the computer screen, will show an icon symbolizing the unmanned devices. A map of a given area will be visualized, along with the units present in that particular area. After selecting a particular platform (hovering the mouse on it) the relevant data will be displayed (current geographical location, technical platforms, photos taken during the mission). The map module will allow:

- Presenting current position of a vehicle or group of vehicles,
- Visualizing an unmanned object, together with its parameters on an on-line map and reproducing the selected unmanned object trajectory together with the detailed flight parameters

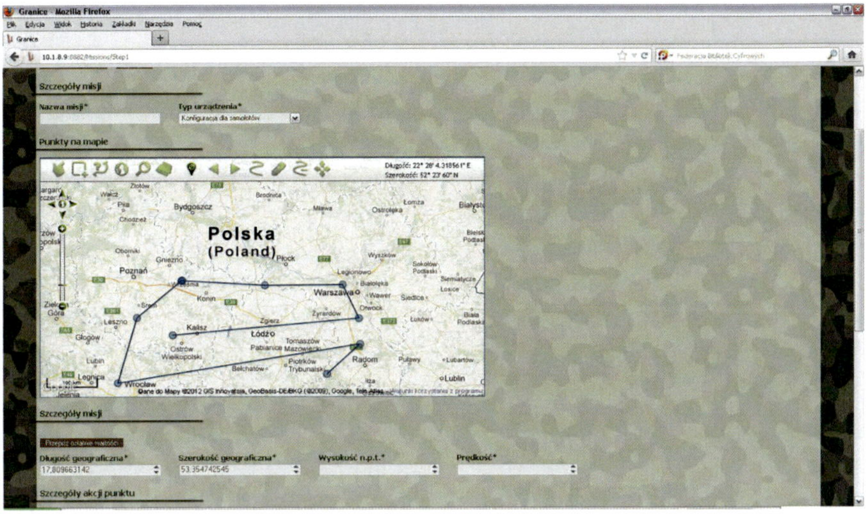

Fig. 1 GRANICE Map Module

- Analysing the data and presenting the results of such analyses in a form of interactive maps, obtaining information about objects from the map, searching for information in the map, connecting map segments to the information contained in the forms,
- Data analysis and presentation of the results of such analyses in a form of interactive maps that will show the relationships between the real objects, allowing to make the most informed decisions, supported by a full review of the situation.

The user functions include: viewing, setting the centre of the map to a specific coordinate, changing the scale, getting the coordinates of a point on the map, updating the position of the selected icon, or adding comments to objects on the map.

3 Control Module

To allow reviewing the basic parameters of a flying object and developing mission scenarios, the control module was introduced. The control module includes a module for recording the flying objects. In this application section, all data will be collected allowing to record the unmanned flying platforms for execution of autonomous reconnaissance and monitoring missions. With the module, it will be possible to document, collect and develop basic information on the equipment and monitoring equipment deployment. Thoroughly developed records will also enable the planned policy of monitoring and controlling borders, preparing detailed reports and gathering information about the areas being monitored. The main objective of the module for recording flying objects is the collection of data, sorting it according

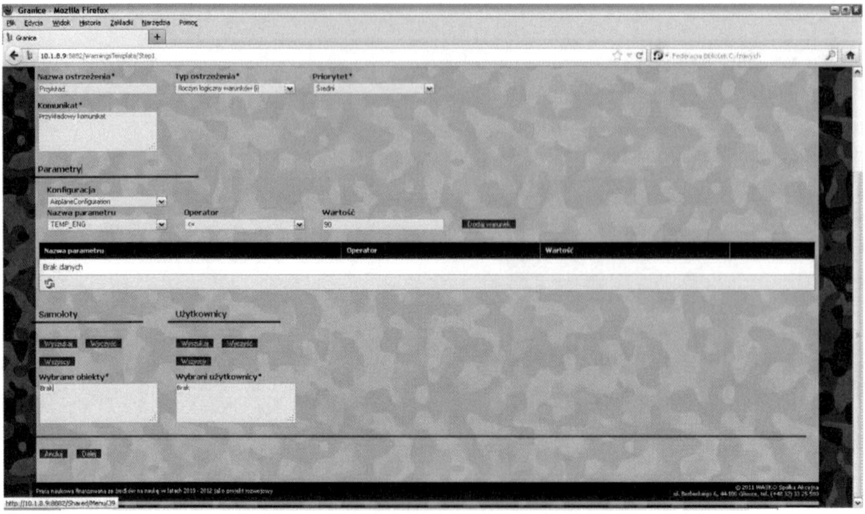

Fig. 2 GRANICE Map Module

to the available filters and the ability to print it in a fixed order. The data recorded will included the following information:

- Technical equipment i.e.: wing area, span, own and overall weight, maximum speed,
- Regarding the equipment, i.e.: registration (from the card, label control, etc.), supplies (consumables, a description of the engine and batteries, extra equipment), service (inspections, technical tests, repairs), a review of fuel consumption (with the possibility to print specific device performance, by the date of execution of the mission), reports and statements to allow the analysis of registration and operating parameters, data from the device assignment to the area (with the possibility of assignment to another area),
- Regarding the details of missions already completed; The history will take the form of a tabular list with paging all registered routes, divided into units and patrols. It will be possible to filter objects by their names, specific mission or date. It will also be possible to make a report for a given period of time,
- Regarding the service, allowing the function for filing repair orders by filling a form (giving the object number and a description of observed failures).

4 Virtual Cockpit

With the virtual cockpit, it will be possible to control the object in real time, to plan a route to of the machine, view photos taken during a particular mission and control of all the technical parameters of the object in flight. Due to the operating conditions it was decided that the most appropriate environment for introduction of the virtual cockpit will be the so called "fat client" architecture. This environment refers to the user's workstation, which is equipped with a set of peripherals, with an installed operating system and a set of necessary applications. Utility programs will be executed out directly and autonomously in the station processing and exchanging data with the user and other computers on the network. All data is stored on the server side.

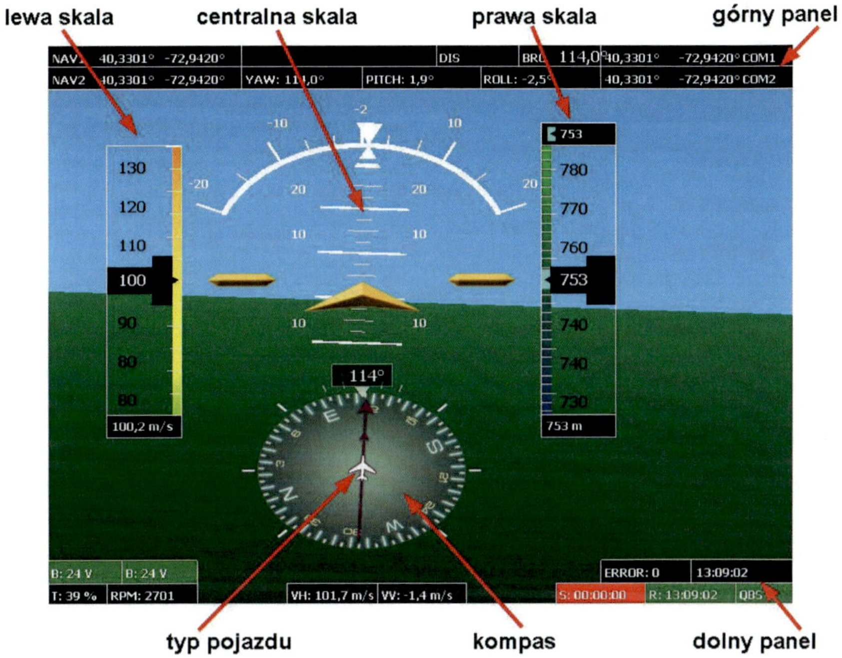

Fig. 3 GRANICE Virtual Cocpit Main Aplication Window

When working with virtual cockpit, a very important issue, which requires attention, is the possibility to divide the resulting image on multiple monitors, to allow closer monitoring of the mission being carried out and the reaction time, allowing for rapid implementation of any corrective action, or controlling the work. Therefore, that discredited the "thin client" architecture, whose main drawback is the limited functionality, due to the protocol used, and delayed interaction with the user, associated with network delays. On the screen of the virtual cockpit, the technical parameters of the facility and flight parameters are presented in tabular form. Also the time needed to reach the destination, while maintaining flight parameters, will be given. The application makes it also possible to transmit the image received from the vehicle to a mobile device. Each vehicle-control panel operator will be authorised to such transmission. The function will be useful when patrolling:

- the area in charge - from the entire unit coordinator level'
- all area in charge - from a superior unit level.

Owing to that, the user directly involved in the mission will be able to view in field what the camera has recorded. To allow possible correction of the patrol group duties, it will be possible to give simple commands by the operator: "Zoom in", "Retract", "Fly to the point", "Suspend" to the aircraft. Thanks to the return channel between the device and the system, it will also be possible to carry out monitoring of the actual position of the mobile device selected on the map online, by the operator in the guards office.

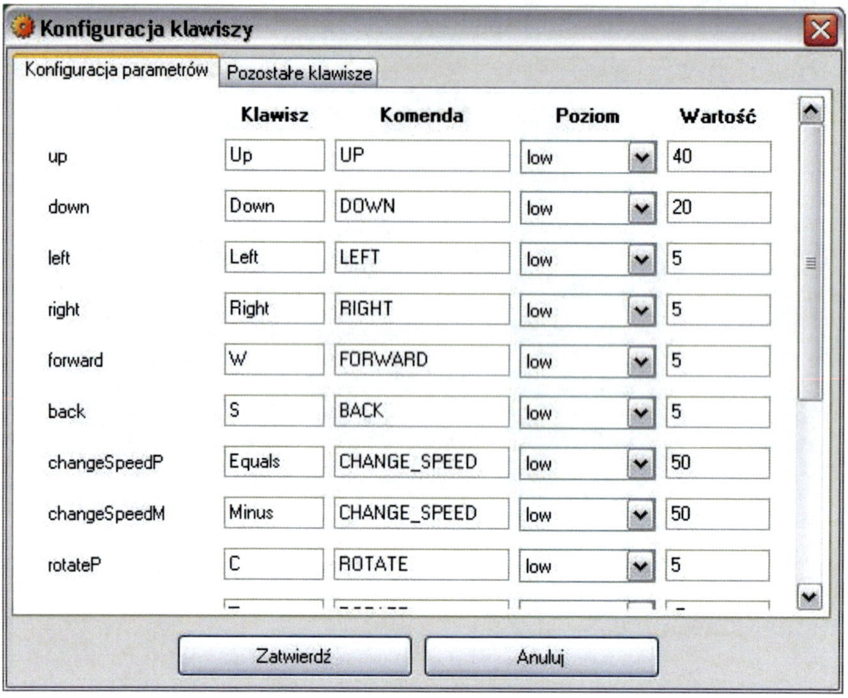

Fig. 4 GRANICE Virtual Cocpit Paramets Configuration

5 Communication Module

The communication module is responsible for the distribution of telemetry data and commands between the computer system and the object. According to the criteria imposed on the system, all platforms will be controlled and monitored by the operator panels, which will be held by each patrol of the Border Guard. At the same time each unit of the Guard will be subject to inspection and monitoring of a superior unit. This will allow to monitor and possibly correct the work service platform. Communication between the operator terminal and the flying machine will be controlled by the monitoring server. This server will be located at any location of the Border Guard. Each foot or mobile patrol, located on the ground will hold of a small, portable computer. Such a device will allow to continuously monitor the flying object and provide the view from the cameras and sensors placed on it. Since the communication between the server and the monitoring operator terminal will be ensured by radio, an additional way to write data from the mission in case of loss of radio communication will have to be considered. Such solution will prevent from a possible loss of connectivity and lack of information about the current location of the object. All information collected during patrols will be additionally stored on an SD card, which will be the standard equipment on every platform. Such card will be pulled out from a flying machine and inserted in a reader located at the operator position. Card data will be imported into the system using an appropriate form. The SD cards used will have sufficient capacity to store data that require large amounts of memory, writing speed, the ability to read and write data. The monitoring server, on the other hand, will have an appropriate port to receive data from the card, the card identification mode, the data transmission and reception mode. It was also planned to create a on-board system to analyse video streams and thermal images to detect living creatures in normal visibility conditions, or under of total or partial lack of visibility, but that will depend on the existing parameters of the link. An important issue for the communication module will also be the telemetry. This solution will enable the monitoring of control and measurement data of distant objects, often located in remote areas. It will consist in placing, in the field, devices that measure the desired value and automatic data transmission by radio or telephone to the headquarters. The control and measurement telemetry systems will ensure the following functions:

- Recording of measurement results,
- Calculating the average value and sharing parameters related with the accuracy of values being read,
- Evaluation of measurement results, formulation of indicators and their ordering,
- Generating signals for the adjustments, e.g. warning signals,
- Providing data about the current value being measured or controlled.

Thanks to that, all images captured during the flight will be grouped and stored. For ease of operation, especially in adverse weather conditions (flying in clouds, in fog or at night), also the called. artificial horizon was introduced to the project. Artificial horizon as a gyroscopic aviation instrument, used to determine the spatial orientation of an aircraft relative to the local horizon plane (angle of tilt and roll). On the face of the instrument, there will be the silhouette of the aircraft, with the horizon - moving relative to each other. The artificial horizon will indicate the actual orientation of the mobile device in space.

6 Video Module

The video module analysing the video and image processing will be a very important element of an unmanned mobile platform. The purpose of vision systems will be to communicate current information about the environment of the object and changes occurring in the same device to the workstation. The functions of the vision system will include identification and location of objects, together with determining their location and spatial orientation, determining relations between objects being monitored, navigation and controlling the flying machine. The functional requirements of a general purpose vision system require:

- Determining the surface and object shapes,
- Dividing the scene into coherent parts that may be viewed, analysed and identified individually.

The software for motion analysis based on visual information will also offer a series of algorithms allowing to analyse motion in the image being recorded. Those algorithms very precisely track the objects recorded, basing on:

- characteristic points,
- shape coefficients,
- colour and contrast distribution.

The software will allow to measure acceleration, speed and movement of the target in a user-specified system of coordinates, and to analyse motion in 3D space (using at least two synchronized cameras).

7 Summary

In terms of works related to the software necessary for the operation. monitoring and remote management of an object, software will be produced for encrypted, wireless data exchange between the object and the ground station, allowing remote operation and viewing of most of the sensor systems on board. It is assumed that the IT system, due to the quantity of data being processed and stored, is a very interesting material for analysis. Therefore, the system will be added an analytical module, built of data warehouses, analytical and reporting software and a server to share and distribute reports. Under the project, a data warehouse will be prepared, which will share data for the analytic and reporting software, and a reporting server, in which the reports or analyses will be located. The reports located on the server will be available via www with access authorised to appropriate users.

References

1. Litwin, L., Myrda, G.: Systemy Informacji Geograficznej. Zarządzanie danymi przestrzennymi w GIS, SIP, SIT, LIS, HELION, Gliwice (2005)
2. Banaszyk, P., Fimińska-Banaszyk, R.: Podstawy organizacji i zarządzania. Wydawnictwo Wyższej Hanzeatyckiej Szkoły Zarządzania, Słupsk (1997)
3. Lemmens, R.: Semantic interoperability of distributed geo-services, Delft (2006)
4. Brodie, M.: The promise of distributed computing and the challenges of legacy information systems. In: Proceedings of the 10th British National Conference on Databases, Heidelberg, FRG (1992)

Architecture and Functional Division of an IT System GRANICE

Marcin Śleziona

Abstract. This paper presents software and hardware architecture and functional division of IT system GRANICE. First part describes used software technology such as ASP.NET, WFC, GIS, ESB, RDBMS, Video Streaming. Second part describes functional division of system. Third part describes communication platform between UAV and IT System. Fourth part presents proposed hardware architecture for system GRANICE.

Keywords: system architecture, GIS, video streaming.

1 Introduction

Effective use of unmanned flying platform requires development of an IT system which will enable collection, processing and visualization of data gathered by it. It is also necessary to enable platform targeting and monitoring the status of its sensors. Data collected from a flying object have to be immediately presented to staff terminals and archived in the central database. The system produced for the needs of this project will enable implementation of all these tasks.

2 Applied Technologies

The information system for GRANICE project purposes is developed in Microsoft.NET technology. There are also used elements of GIS, ESB, video streaming and author's solutions for data exchange between modules. Applications

Marcin Śleziona
WASKO S.A., Berbeckiego 6, 44-100 Gliwice, Poland
e-mail: `m.sleziona@wasko.pl`

A. Nawrat et al. (Eds.): Advanced Technologies for Intelligent Systems, SCI 440, pp. 185–194.
springerlink.com

included in the system are developed in Web technologies and fat client. This scope of used technologies allows to develop a comprehensive and scalable system, but it brings with it challenges of research and development related to reliability and performance of the solution.

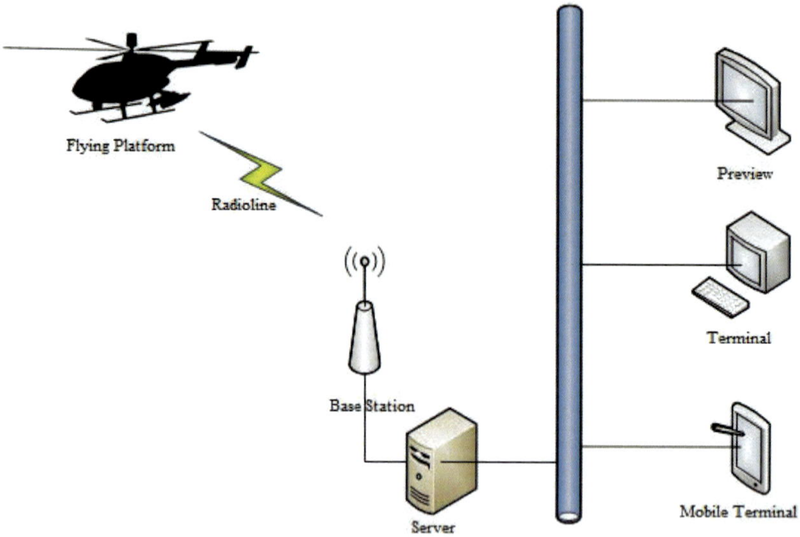

Fig. 1 GRANICE system architecture

Web Application which is the main part of GRANICE IT system was developed in Microsoft Visual Studio 2010 using ASP.NET Model View Controller 3 design pattern. This pattern divides the application into the following layers:

- model, which is a representation of the application logic,
- view, which describes how a fragment of a model is displayed,
- controller, which receives input data from a user and a person managing view.

Use of such a design pattern allows separating logic of application from presentation layer. There is also important MVC3 framework used in the programming environment. It forces a logical division of applications and elaborating applications according to the model which ensures high reliability and system performance. An important element of any application, especially Web one is its user interface. Standard components designed for building Web applications in which Microsoft Visual Studio environment is equipped provide only basic features, and their extension with additional elements is labour consuming and it entails the risk of mistakes in implementation. Therefore, on the market a review was carried out of components allowing for construction of an ergonomic user interface in ASP.NET. As a result of the analysis of the available solutions, Telerik components were chosen. Selected technologies enable development of efficient and scalable Web application being an element integrating other domain applications included in the system. The system expansion with additional elements is also possible and does not require considerable work effort related to the integration. Applications based on Web technologies have also some limitations, such as refreshing rate of content, limitations of the user interface, and problems with communication with peripheral devices. Therefore it was decided that the application which executes a virtual cockpit functions will be carried out as a thick client in Windows Forms technology with the use of DirectX. Such a choice of technology will provide easy access to object control devices (yoke) and allow for smooth representation of flight data and artificial horizon. A flying platform will be equipped with cameras for monitoring of a selected area of land. The signal to the ground station is transmitted in analog form, but to allow its archivisation and transfer to service terminals conversion into digital form is necessary. The conversion is executed in hardware frame grabbers used to convert analog PAL to digital MPEG-4. The digital signal is received by QBService and transmitted to streaming Unreal Media Server. Streaming Server archives in a dedicated repository and provides a picture for all client stations. A solution based on one central streaming server makes possible to control fully the video signal and minimizes the size of transmitted data. The applied streaming server allows you to upload an image on links with very low bandwidth. The standard-defined profiles allow streaming to the following links:

- Dial-up (40 [kb / s])
- Slow DSL (150 [kb / s])
- DSL (256 [kb / s])
- T1 (400 [kb / s])
- T3 (600 [kb / s])
- LAN (3000 [kb / s])

An image can be streamed in many different formats. Supported are Unreal Streaming Media (UMS), Microsoft Media Streaming (MMS) and Real Time Messaging Protocol (RTMP). An important parameter is the delay occurring between accepting an analog input image and presenting it on the terminal. In case of applied solution it is possible to obtain the following delay times:

- 0.05 - 0.01 [s] - for calls within the LAN,
- - 0.3 [s] - for others.

To present an object on a digital map GIS technologies from ESRI and others companies were used. The elaborated solution was necessary for the development of two independent mapping applications. The first of these makes it possible to show an object on the map in the web application. For this purpose a dedicated map server shared by all client terminals was used. The second one designed for the mobile terminal, uses ArcGIS the runtime 9.3 and has been implemented in the form of a fat client. This division will allow minimizing exchange of movement between the mobile terminal and the system and support mapping underlay in standard formats. The information system GRANICE enables you to work on multiple terminals and in multiple locations, it is also required to display up-to-date information (with at least 3 frequency) [Hz] of objects on each terminal. There were tests carried out which encompassed transmitting data through the central database based on Microsoft SQL Server. These tests clearly confirmed that the relational database does not fulfil the requirements. It was impossible to obtain the required frequency of data refreshing. Therefore, it was decided to develop its own mechanism for transmission of telemetric data from the application object to the terminals. QBService application was built on the basis of Windows Communication Foundation technology. WFC integrates and unifies all the existing Microsoft technologies such as:

- .NET Remoting,
- MSMQ,
- COM+ (EnterpriseServices),
- WebServices.

Thanks to the use of WFC application QBService supports various communication protocols, various transmission media, and different sets of measurement data and control parameters. It enables sending messages between the object and field applications, as well as commands from an application to an object, moreover it is also fully configurable and has a powerful mechanism for authentication. The most important application features include:

- monitoring and management of an object,
- easy integration with external systems,
- it enables easy upgrade with new function blocks without interfering in the kernel and recompilation necessity.

The selection of a relational database for the system was guided by the following criteria:

- performance and scalability,
- support for data service and geographical features,
- the possibility of integration with external components,
- technical support,
- licensing method.

These requirements were satisfied by PostgreSQL 9, Oracle 11g and Microsoft SQL Server 2008R2 databases. Due to simplicity of integration, consistent programming environment, performance and safety it was decided to choose Microsoft SQL Server. The information system GRANICE is to provide the data collected in the central database through the ESB to other information systems. An analysis of the available solutions on the market-type ESB was made. Their compatibility with already selected technologies and technical capabilities were taken into consideration. As a result of this analysis it was decided to use Microsoft BizTalk Server 2010. It allows easy and efficient integration with Microsoft SQL Server database and meets all requirements of the ESB in GRANICE project. The above selection of technology allows for developing of a coherent system, which shall also will be relatively easy to use and maintain.

3 The Functional Scope of the IT System

GRANICE IT system includes comprehensive management and direction of flying unmanned platforms. It is designed for multiple groups of users performing their tasks, both in headquarters units and field databases. It forces a division of application into functional modules for tasks of each user group. The basic element of the system is web application that enables carrying out duties associated with supporting facilities. The application has been divided into the following functional modules:

- administration module and user authorization,
- records module,
- maps module,
- video module,
- reporting module.

The administration module and user authentication functionalities encompass functionalities connected user management and general and administrative functions. It is available to add users and roles. For each account will be possible to attach multiple roles. User Role defines the scope of access to system functions. This module is also responsible for user authentication. Login to the system takes place once, i.e., the transition to a different Web application module does not require re-entering the password. The records module collects information about objects. Each object can have its own configuration for the exchange of data with the system. In the records module there is also information such as:

- object name,
- belonging to the organizational unit,
- registration data,
- carried out inspections and service repairs,
- machine specifications:
  - the bearing surface,
  - the span,
  - the empty weight,
  - the total weight,
  - maximum speed.

In addition to the use of the system on fixed positions it is also possible to obtain basic information about an object on a mobile terminal. The scope of functionality of the mobile terminal includes presentation of the up-to-date position of an object on a digital map and presentation of an image of the cameras. The mobile terminal application has also the possibility of issuing basic commands. These commands are divided into two groups. The first concerns camera head control commands. A user of a mobile terminal may rotate the camera and make close-ups. The second group of commands concerns movement of an object. These include only basic commands such as hovering, stop and the continuation of mission, emergency landing. Mobile application is equipped with self-reliant digital map independent from web application. This map enables user to take advantage of different underlays of map depending on the type and nature of the mission of the platform. Mobile application was made in such a way as to minimize the amount of transferred data. In most cases, management of the platform in the selected area will be carried out by indicating a point on the map. It should, however, be assumed that there may occur cases in which this form of control over the object will be insufficient. Therefore it was decided to add to the system an application which maps a virtual cockpit which, together with the application mapping and image from the cameras will take full advantage of the flying platform. Virtual cockpit was implemented in the form of so-called fat client. The application aims to provide primary flight characteristics of the object and enable control of the object in real time. The main element of the application is an artificial horizon.

The application provides the following data on graphical components such as:

- artificial horizon,
- coordinates,
- course (degrees),
- velocity relative to air,
- rate of ascent / descent,
- height,
- value of the throttle,
- engine rpm,
- control informing about the connection with QBService,
- control informing about receiving control data (the colour control will depend on the delay),
- current time, the time of receiving the last frame.

In addition to the above data, the application enables presentation of any data downloaded from the QBS in text form. For each of the presented values may be defined auxiliary parameters such as:

- the minimum value,
- the maximum value,
- optimum range of values,
- name as well as abbreviation of the parameter.

Depending on these parameters the value will be shown in different colours. If the system does not receive data of the parameters or they are outside the defined range there will be displayed an error message concerning parameter. The component will allow to disable each element separately.

Virtual cockpit also allows control of the vehicle. For this purpose, the software was integrated with hardware yoke. Each of the values taken from the yoke can be calibrated and assigned to the corresponding axis of the controller. Data sent from the virtual cockpit to the flying platform are archived by QBService to a central database. Control commands sent to the object are in a simplified form. These commands define the direction in which you want to move the object, change the flight altitude, rotation in a certain direction. Commands are interpreted and decomposed at the level of a flying object. A user operating the virtual cockpit will not be able to choose the speed freely at which the object will move.

4 QBService Communication Module

All modules use the central database; however, due to the nature of the system it was necessary to develop a mechanism for exchanging telemetry data and the commands between each of the system modules and the subject. The developed solution allows controlling and receiving telemetry data from many flying objects. The collected data are archived in a central database and sent to the appropriate system modules. The communication module also receives commands from the system and transmits them to flying objects. In the application QBService the flying platform is the couple: flying unmanned platform and the agent application that with the use of native protocols communicates with a definite device. The communication protocol between the QBService application and the agent applications is the same regardless of the type of flying machine, which it controls. Monitoring applications and controlling applications (or in a particular case monitoring - controlling ones), subscribe to QBService application to monitor or control parameters of specific flying machines. Through QBService applications appropriately receive up-to-date values of monitored parameters or send to the object the sent control vectors. The main task of QBService application is to service other elements of the flying machine control system. Therefore, the key tasks carried out by the software are happening "in the background" and they are not visualized. Only the configuration layer, which allows for actual work and cooperation of QBService application, is performed in a graphical way and it is supported by the GUI in Web technologies. Full system, that allows controlling the flying device, consists of several independent software components that have to cooperate and communicate with each other in an appropriate way. For the purposes of further deliberations, the elements of the system will be divided into four blocks.

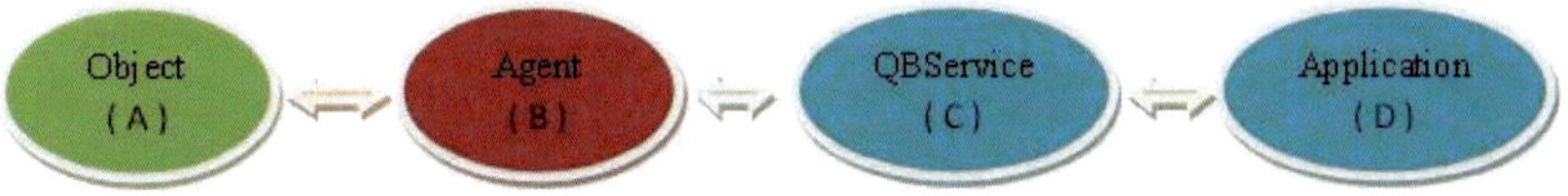

Fig. 2 Communication between the blocks

- Block A - flying platform equipped with a communication interface which allows to monitor the platform parameters and to influence its state. Together with the block B they form "object".
- Block B - Agent. The software that implements the communication protocol for a specific device A for entering data into the block C. The block B has no additional functionalities. Variables recorded by the agents are strictly defined and have the status of input parameters (state vector) and output (controls vector). State vector can be modified by the agent after receiving information from the object, controls vector is sent by the applications of block D. Together with the block A they form "object".
- Block C - QBService Software. It is a "warehouse" which stores data. It stores all the parameters and variables collected in a central database.
- Block D - External Applications. Applications that monitor, steer, or monitoring - steering. Diffused character of the system design allows to present data in a variety of unrelated applications. These applications have a common data exchange protocol with block C.

QBService application was developed to allow flexible and fully customizable creation of connections between the elements associated with the transmission of data. Hereafter are shown the possible connections.

Connection A-B

Agent is a hardware-dependent layer and therefore it is impossible to create this part of the system as a universal solution. Agent is the layer that is created independently for each platform (communication protocol) which is to be connected to the system. Connection A-B is implemented in an independent way from the QBService application. There are three possibilities of configuration of device connection with the agent:

- one agent - one device
- one agent - many objects (the situation may take place in case of the monitoring of several devices with the same communication protocol)
- many agents - one object (the situation occurs when an object has several communication protocols).

Connection B-C

B-C connection is a connection between the agent and QBService. Communication between these blocks was based on WCF technology and is universal for all agents. Communication can take place between any number of agents (type B blocks) that support any number of flying machines with one QBService application.

Connection C-D

Communication between block C and block D is based on the observer-observed pattern. Software D "interested in" monitoring a particular variable or group of variables of a selected device registers itself as an observer and block C notifies it when there is a change in the observed set. Software from block D can freely modify the vectors of control devices, which will automatically call the appropriate procedure for agents.

5 The Architecture of Hardware

The appropriate hardware infrastructure was chosen for developed software to ensure correct performance and reliability. The system was divided into the following environment: " AP1 Application (Web Server - Microsoft Windows Web Server with IIS)

- AP2 Application (Microsoft BizTalk Server 2010 Standard and communication module)
- AP3 Application (Server map)
- BD1 database (Microsoft SQL Server 2008)

All the environments work under the control of Microsoft Windows Server 2008 R2 in the VMWare environment and share the infrastructure of backup system and HP Blade physical server infrastructure. Environments were built with a single blade server from HP, based on a 12-core AMD processor, which allows the use of virtualization without the necessity to incur compromise between performance and functionality. For both hardware and software redundant solutions were used according to the requirements. Taking into account the high availability requirement, the selected HA (high-availibity) solution allows, where appropriate, doubling each functional component, starting from the virtual machine ending at active network devices and the components responsible for supply and cooling. HP blade servers are equipped with built-in network cards with 10GbE with Flex-10 technology, which allows dividing the physical interface into 4 virtual ones, which are visible to operating systems as independent network cards. This solution also enables programmed speed control obtained through various virtual network cards, which fit into the total bandwidth of 10GbE. Running virtualization software or operating systems can take place via the SAN network or internal drives in RAID1 configuration. Application servers and databases were virtualized by VMware vSphere 4. Virtualization layer is a system solution that is installed directly into physical hardware. The solution has the ability to support multiple instances of operating systems on one physical server. It is characterized by the maximum degree of hardware consolidation. The software allows you to configure virtual machines with access to 255GB of memory and to configure virtual machines 1, 2, 3, 4, 5, 6, 7 and 8 processors. The virtualization platform uses up to 12 physical processor cores. It is largely independent from the hardware manufacturer. It is possible to develop the infrastructure with additional services in an easy and quick way without compromising performance and availability of other services by adding more servers. VMWare vSphere 4 supports, among others, the following operating systems: Windows XP, Windows Vista, Windows NT, Windows 2000, Windows Server 2003, Windows Server 2008, SLES 11, SLES 10, SLES9, SLES8, Ubuntu 7.04, RHEL 5, RHEL 4, RHEL3 , RHEL 2.1, Solaris version 10 for x86, NetWare 6.5, NetWare 6.0, NetWare 6.1, Debian,

CentOS, FreeBSD, Asianux, Ubuntu 7.04, SCO OpenServer, SCO Unixware. It allows assigning more RAM to virtual machines than physical RAM server resources and virtual machine access to more disk resources than physically reserved. The use of the physical virtual infrastructure is monitored. It is possible to snapshot instances of operating systems for the needs of backup copies without interrupting their work, and cloning operating systems along with their full configuration and data. Backup system was built on a dedicated server blade and tape library systems Quantum Scalar i500. The tape library is equipped with two tape drives connected to the LTO5 backup server through the SAN network. In addition to the backup server space has been made available with one disk array, to make copies D2D2T (disk-to-disk-to-tape). On the backup server and virtual machines Symantec NetBackup 7.0.1 software was installed which is responsible for the backups.

6 Summary

Selected technologies and their use comply with an information system that will allow efficient use of flying platforms. In addition, it will be possible develop further of the system and exchange data with other IT systems. The exchange of these data can be done through Microsoft Biztalk as well as downloading telemetry data directly from the QBService communication module. Selected hardware solutions enable development of infrastructure alongside with development of software and data stored in it.

References

1. Evjen, B., Hanselman, S., Rader, D.: ASP.NET 3.5 z wykorzystaniem C# i VB. Zaawansowane programowanie, Helion, Gliwice (2010)
2. Johnson G.: Accessing Data with Microsoft .NET Framework 4. Microsoft (2010)
3. ESRI ArcGIS, `http://www.esri.com/software/arcgis/index.html` (visited August 15, 2011)
4. Litwin, L., Myrda, G.: Systemy Informacji Geograficznej. Zarządzanie danymi przestrzennymi w GIS, SIP, SIT, LIS, Helion, Gliwice (2005)
5. Microsoft SQL Server, `http://www.microsoft.com/sqlserver/en/us/default.aspx` (visited August 15, 2011)
6. Cirrus SR22 Integrated Avionics System Pilot's Guide. Cirrus Perspective by Garmin (2008)
7. UnrealMedia Server, `http://www.umediaserver.net/umediaserver/index.html` (visited August 15, 2011)

Integration of IT Systems with the Use of Microsoft BizTalk Server Platform

Tomasz Krawczyk

Abstract. The paper presents an approach to designing complex IT systems based on the EAI architecture. Enterprise Application Integration (EAI) is an integration framework composed of a collection of technologies, methods, and tools aimed at modernizing, consolidating, and coordinating IT systems. The work also gives a description of the possibility of using Biztalk Server as the software platform to implement systems integration infrastructure. An example of a system that can provide an important part of complex integrated system is the GRANICE system. Although the GRANICE system is fully functional environment, using the integration platform which is an intermediary in the communication with other systems,we can significantly extend its functionality.

Keywords: integration of IT systems, service-oriented architecture, enterprise application integration, Biztalk server, message, pipeline, orchestration, system GRANICE.

1 Introduction

Having comprehensive precise and sufficient information significantly determines the relevance and effectiveness of decision-making, understood as making an informed choice between different variants of problem solving. The information has to contain data processed into understandable form for people who use it in order to be effectively used. The use of information depends largely on possible transmission among particular entities. Nowadays, complex IT systems are responsible for collecting and transforming data set into information [0]. Usually a given organization uses a few or several different information systems and services of external

Tomasz Krawczyk
WASKO S.A., Berbeckiego 6, 44-100 Gliwice, Poland
e-mail: t.krawczyk@wasko.pl

A. Nawrat et al. (Eds.): Advanced Technologies for Intelligent Systems, SCI 440, pp. 195–210.
springerlink.com

companies. Each of these systems is responsible for processing data covering only a dedicated area which is a part of the activities of the organization. Very frequently encountered is a situation in which some data are duplicated across multiple systems, which in turn leads to the fact that these data are inconsistent (not sure which system stores the up-to-date data). Usually applications being an essential part of an IT system are located on different operating systems, they use different database engines, utilize different communication protocols for communication, they use different data structures characteristical for specifications of its operations and are developed in different technologies. From the perspective of the end user, it does not matter which technology was used to collect and transform data, he expects the information that will allow him to make the right decision. In most cases in order for information to carry useful content it has to come from several sources and be processed by many IT systems. Sometimes it is necessary in order to supplement information to refer to external data sources or the use of external systems (not being a part of the IT structure of the organization) to transform the information into the form desired by the user. Several years ago preparing by only one IT system expected and useful data, (which usually was designed to process information from only a dedicated area) required more and more time, which resulted in substantial increase in costs. With the increasing requirements as to information systems provided by users, the closure of the functionality in one system became more and more troublesome. Besides, it was noted that chosen functionalities of various systems fulfilled the same tasks and that a similar task is carried out by some systems more efficiently than others. Companies that specialize in providing dedicated solutions appeared.

2 Service-Oriented Architecture and Integration Techniques

The increasing challenge posed to information systems concerning the delivery of information meant that there was a need and indeed a necessity to ensure cooperation between different IT systems, both within a single organization and among organizations. This cooperation included both data exchange and the use of services offered by other systems for data processing. It became necessary to create infrastructure of an IT system in which different systems work together - therefore a need occurred to integrate systems. There also appeared the concept of distributed systems, consisting of independent and cooperating components, which are characterized by transparency. As part of such environment the data processing takes place in different systems, but from the perspective of the end-user this environment is seen as a consistent, complete entirety. The process of transformation data under distributed computer system is directly related so-called distributed processing, that is, information processing, in which discrete components (both hardware and software) may be deployed in different locations (geographical and / or organizational) [5] [2]. In the overall assessment the key features of a good distributed computer system are the following:

- Openness to extend and adding functionalities by attaching new elements,
- " Scalability - the possibility of using the services offered by a growing group of users without significant loss of performance.

One of the first approaches to integration of applications in distributed systems was architecture based on point-to-point, meaning direct connecting of components. This solution requires a thorough knowledge of integrated systems, i.e., their communication protocols and data structures. For each of the two systems there is a need for preparation of a separate application which is responsible for the exchange of information between these systems. The situation becomes more complicated if we want to integrate more than two applications. For n systems, the number of such connections amounted n x (n-1) /2 in the worst case and twice as many interfaces, since each pair of A-B applications required creation of two interfaces (one responsible for communicating with the application A, using communications protocol for this application and the second one for communication with the application B) [2][12].

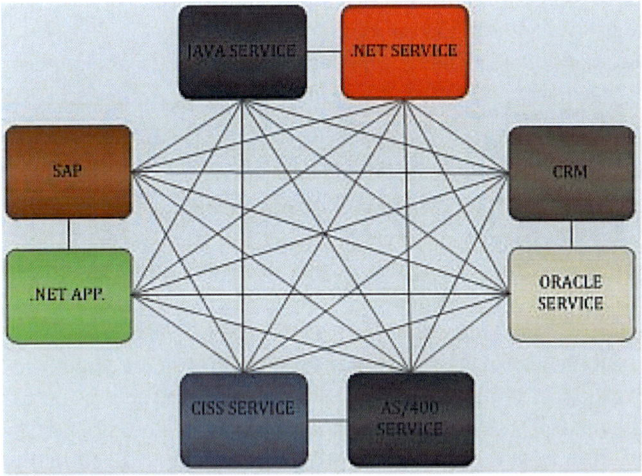

Fig. 1 Point-to-point architecture

The solution according to the original assumptions met its expectations, however, rising expectations of users and the need for cooperation with an increasing number of systems has not proved well. Currently placed demands impose for created distributed systems the need to use already existing components without the necessity to create a whole system from the start. Individual components of a complex system running on different platforms should be treated as indivisible components, which are the basis to create a business process. This approach forces heterogeneity, i.e. varied size of data structures, different functionality, different data models in place of harmonization or unification of all components, tools and communication protocols. Distributed systems are naturally heterogeneous. The creator of the system usually must use components with different properties performing different tasks on different platforms or using different data structures. The system consists of a number of tools: mainframe stations, SAP hosts, databases, services or applications [2]. Changing requirements make the need to implement a new element, different from hitherto ones. During this operation, new business functions are added and components are changed. This is possible if the underlying architecture tolerates heterogeneity and allows for flexible management - architecture point-to-point does not meet these requirements [2]. Problems associated with the capability (e.g., services) of entering the run-time interaction of two software artifacts with the aim of reaching common objectives (interoperability) appear when we want cooperation of services or data sets that are heterogeneous [3].

We can distinguish several levels of heterogeneity which result from the need of different IT systems to communicate within the distributed system. These include [7]:

- heterogeneity on the syntactic level such as conflict in encoding characters or serialization of data set,
- heterogeneity on the structural level - which results from the use of different types of representation of information about the same object,
- heterogeneity on the semantic level - due to discrepancies in the expected meaning of concepts used.

The modern process of integration assumes acceptance of the strategies, in which already existing systems will be an integral part of the newly created distributed system. Often, it forces solution to problems arising from heterogeneity on syntactic and structural level, and it leads to problems with stability and backwards compatibility. In response to new requirements for cooperation between different systems there came the concept of creating IT systems, which puts the emphasis on defining services - SOA (Service Oriented Architecture) for creating business processes with different owners. SOA based concept accepts heterogeneity of systems and places special emphasis on flexibility in implementation, which should be focused on the possibility of continuous changes, extensions and updates [2]. SOA does not constitute specific architecture. It is based on the assumption that business functionality is included in services. Focusing on the business value of interface, the service creates a bridge between business and IT environments. In the SOA paradigm we can distinguish three main concepts [6]:

- SOA Service aims at separating the business aspects of the problem from the IT software representation and placing implementation of this aspect in the service. Technical details should not be visible outside. Only the interface understood for business environment should be exposed,
- Interoperability - SOA shall enable interaction between two software artifacts, which in order to achieve their goals have to communicate with each other during the run,
- Loose connections - SOA reduce the links between systems, thus limiting the modifications or errors and greatly increase flexibility.

In SOA, we can also distinguish three main components, which include [6]:

- Infrastructure,
- Architecture,
- Processes.

Infrastructure defines the tools to achieve high interoperability, such as transformation (transformation of data), routing, service management or security provision. The architecture is necessary to reduce the specific implementations of SOA, so that they become functioning managed systems. The processes are responsible for the realization of tasks. The process involves the concept of business process modelling i.e. preparation and defining of services that constitute a business process. Service-oriented architecture proposes the implementation of business products through the execution by the instrumentation as atomic as possible and reusable services which are technological beings. In the mature SOA architecture service created for a new business product is a complex service benefiting from already existing services, and its execution time is relatively short [6]. One of the potential implementations of technical demands of SOA are Web Services. However, Network Services do not meet all relevant demands of SOA, these are in particular guarantees of interoperability and assurance of sufficient level of loose connections. In recent years there have been new concepts concerning construction of distributed systems, so that they met the assumptions proposed by SOA. In the most important ones we can include the systems based on Enterprise Application Integration and Enterprise Service Bus. EAI should be understood as middleware type application software (platform) to create architecture integrating heterogeneous systems, by ensuring the feasibility of business flows. It is based on a centralized the hub-and-spoke architecture, in which all communication takes place via a central hub. This solution enables data sharing between multiple information systems and automation of distributed business processes within the company [2]. In the hub and spoke architecture connections of (integration) applications occurs in the central hub. The connection to the application is done by using so-called adapters (spoke). They are responsible for the connection to a given application as well conversion of downloaded data from the application to data format in which hub is operating and vice versa, i.e. conversion of hub data into the data format used by applications [4].

The basic assumption of ESB assumes that each system connects only with the bus to be able to communicate with each system connected to the bus. ESB postulates the existence of native intelligent transport layer to which any system can be connected by plug-ins.

Services available within ESB are independent from the transport layer (they can be used by Web Service, JMS, MSMQ, etc.). Messages flowing within the ESB can be redirected and modified according to specific rules. Modern bus provides monitoring of operation of services as well as virtualization of services through load balancing, failover, mediation between transport protocols, often it has a BPM solution (Business Process Management) for implementation of long-lasting business processes. [4].

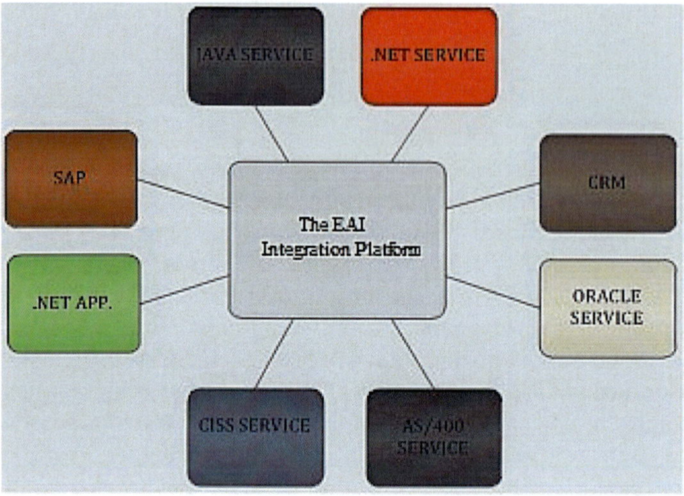

Fig. 2 The EAI Integration Platform

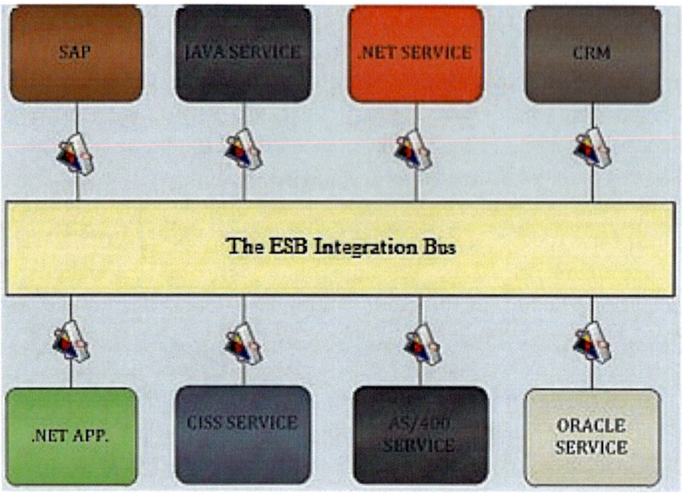

Fig. 3 The ESB architecture

3 Microsoft Biztalk Server - Integration Platform

One of the systems of this type (providing support for the architecture of EAI as well as ESB) is a Microsoft's BizTalk Server commercial product. It provides the infrastructure to the integration of applications as well as it provides an environment to connect and build together any environments, consistent with the vision of SOA. It provides the functionality of the integration of data exchange and integration on the level of business processes. Microsoft BizTalk Server provides environment for designing business processes responsible for the exchange of messages, and it has mechanisms allowing to define the acceptable definition of input and output data and patterns of their transformation. On the one hand it is the tool for connection and integration of previously independent applications which were running along with the establishment of business rules, standards and interfaces of the communication, on the other hand it can serve the same role in relation to entire systems, between which it is necessary to build mechanisms for data exchanges and joint services. Thanks to BizTalk server exchanging information between systems in a complex infrastructure is based on industry standards and architectural patterns such as: (SOA) Enterprise Application Integration (EAI), Business to Business (B2B) Complex Event Processing (CEP) and Enterprise Services Bus (ESB) [10]. The main elements of Microsoft BizTalk Server are Messaging Engines as well as so-called orchestration (Orchestration Engine), which form the basis for integration architecture of data exchange between different IT systems. The role of the messages engine is receiving incoming messages, the analysis of these messages in order to identify their format, content analysis to determine how a message should be processed, the delivery of processed messages to the destination and tracking the status of the processed message. The role of the orchestration engine is to perform the business logic according to the specific business process. Business Process is a set of activities that meet the desired needs of the end user (may take a minute, hour or even several months) [10]. Microsoft BizTalk Server performs tasks with the use of publishing and subscribing model (publish-subscribe). In this model there is used one-way synchronic broadcasting communication mechanism (which in some cases may require the use of an external program of infrastructure Publish / Subscribe). Input data are received by input port, which consists of two parts: an adapter responsible for the reception of data from the source system as well the input stream. Input - receiving stream consists of four elements, whose role is to perform the following tasks: Decode (Decode) - responsible for decrypting or decoding the message, disassemble - the breakdown (Disassemble) - which is transforming the message, checking (Validate) - responsible for checking the correctness message (if the message is the syntactically correct and that it meets the requirements specified by the XSD scheme) and determined the identity (Resove Party) used to establish the identity of the sender (e.g., signature verification if it is defined). The main task is converting the input stream received data into an XML format (the main format used BizTalk platform) - received data have a message form [10].

XML is extensible tag language, platform-independent, which allows easy replacement of documents between heterogeneous systems. Requirements for XML document structure can be determined using the XSD scheme. For the needs of searching documents in XML format XPath has been created, which provides an expression enabling to refer to any elements or attributes of the document satisfying the required conditions. For the transformations of the document described by a single XSD scheme output format compatible with any other scheme the XSL transformation is used [13]. Data received by the adapter and processed by the input stream at a further stage of processing have the form of a message in XML format

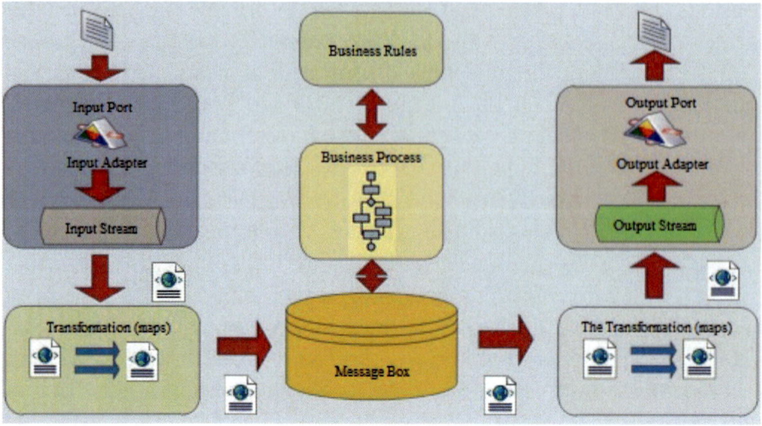

Fig. 4 The idea of the BizTalk Server integration platform

with strictly defined, described structure of the XSD schema. Sometimes it is necessary to change the structure of message to another. Transformations can be simple or complex and may involve combination (data grouping) or require additional calculations. BizTalk Server uses so-called maps for the description of the definition of such transformations. The maps contain business logic required for data transformation between two different schemes. Graphic editor provided by the manufacturer is used for building maps, which comes down to a graphical identification of relationships between elements (or attributes) of XSD schemes that describe converted messages. These relations can be extended with so-called functoids - (there are over 70 availoable as well it is possible to build one's own) i.e. special functions that allow to perform additional operations on data (e.g., a combination of two values of text type). Received and processed messages go to so-called inbox messages. Message box is a database called the MessageBox on the MSSQL Server, which ensures that the data will be sent to the sender - will not be lost in case of accident or error. Stored messages can be transmitted directly to the output port, which consists of the output stream and the adapter. The task of the output stream is to transform data from XML into the format expected by the target system, while the role of the adapter is providing data to the target system. Transmitting - output stream consists of three elements responsible for the following tasks: preparing message to ensure that the message can be created for sending (Pre-assemble); creating messages to send (Assemble) - by performing additional operations, such as the transformation of message from xml format to the flat file format; encoding (encode) - responsible for the encryption or (and) coding the message. The need to perform additional operations on the received message is often, that is the business process execution according to the established workflow. Workflow determines how the message flow between objects involved in the processing and defines what operations are going to be exdcuted at a given stage of processing. To describe the workflow in Microsoft Biztalk Server so-called orchestrations are used, which control the flow of business processes, and the orchestration engine is responsible for their run. Similarly to creating maps orchestration is done by using a dedicated graphical editor [8]. For descriptive purposes of orchestration dedicated BPEL language has been created. It is used for specification of the parties (partners) involved in the processing messages sent among the parties and describes the business logic of the process. In the BPEL language we can distinguish two types of actions: simple and complex. Simple actions are: service call, receiving and replying to queries and appending data to the message. Complex actions include the following structures: sequence, parallel execution, or so-called nondeterministic choice instructions (while, switch and pick) [9]. Environment of orchestration handles all the difficult issues related to implementation of long-term, dynamic processes, such as transactions, compensations, control flow, interconnection of messages. Orchestration engine uses business rules engine, which provides automation of rules and business intelligence, taking over all the tasks associated with sequencing commands and control flow. It is used to define and modify a dynamically changing business rules, written in the form of rules. The rules are expressed in natural language using dictionaries or with the use of describing elements of the rules by means of names easy to understand. A powerful inference mechanism determines the result of the implementation of rules and may deliver other rules so as to reflect a change that has taken place in the environment. The executive mechanism of rule engine finds the need to comply with additional rules depending on the state of the object changed by execution of the previous rule. Within processing orchestration engine retrieves messages from the inbox messages, converts them and re-writes the processed message in messages box, which then goes to the output port [10]. Adapters for BizTalk Server system play a very important role - they are responsible for communication with external systems. They provide support for data streams specific to a given technology (along with support for metadata associated with a given receiving or broadcasting protocol). BizTalk Server provides a large amount of ready-made adapters, which can be divided into adapters connected with given communication standards service (such as MSMQ, HTTP, SMTP, FILE, FTP, WS *, WCF) service of industry protocols (such as HIPAA, HL7, SWIFT) and application adapters - dedicated to communication with given applications (SAP, Siebel, Tibco, Oracle ESB / DB, EDI) and outsourced companies adapters [8]. Additionally, there is a possibility to create your own adapter based on the Adapter Framework. To the main advantages of using BizTalk Server as an integration platform we can include [8]:

- Effective communication both internally and with external entities,
- Automation of communication with other organizations' applications and between applications within the company,
- Ability to define business processes and event management, and fundamental in-process productivity ratios to ensure its effective and efficient action,
- Building dedicated solutions for every sector in any geographical region,
- The possibility of using a standard set of reusable services, which significantly reduces the costs associated with the creation of business processes,

- Modular - an extensible architecture that allows for adding plugins (adapters) of independent producers, facilitate changes in the structure and standardization of message,
- The use of standard - generally used technologies such as XML and Web services,
- The ability to build business processes graphically,
- The ability to monitor business activity - downloading information about the up-to-date state of the process,
- Integration with other Microsoft products such as SharePoint, Office, or products associated with Business Intelligence,
- Extensive programming environment - Visual Studio and Team Server or MS Project,
- Relatively low cost of implementation and maintenance.

Due to its extended possibilities BizTalk Server may be used for the following scenarios [8]:

- Integration of the application on the level of data exchange,
- Automation of processes in which the process requires the use of external services to perform a given activity, which is a part of the process,
- Infrastructure optimization, so that it will be possible in a simple way to join new systems to the existing infrastructure and providing efficient and effective cooperation between them.

4 Integration Capabilities of GRANICE System with External Systems

The system GRANICE consists of two basic elements into which we can include flying devices and software. The IT system established within the project contains many complex components, among which we can distinguish:

- software responsible for communication with a flying object, which includes downloading data from the object and transferring basic control commands,
- software for monitoring objects - including monitoring of primary flight parameters (including parameters related to the behaviour of objects such as fuel consumption, speed, etc., and characteristic parameters for the flight such as location or image from cameras), including applications enabling to visualize these parameters and in the case of data concerning the position showing the current location of an object on the map,
- software to organize and manage missions, including applications enabling to plan the mission in details, including the set operations to be executed on each stage of the mission,
- Software that lets retrieve flight information of an object at given time,

- Software that lets managing authorisations for both objects (through which applications can communicate with the object, and which applications have access to data transmitted from the object),
- A series of small applications that implement other functions (monitoring system activity, etc.).

This system is a complete entirety fully executing the tasks set for it. Due to its specificity and the fact that during performance of tasks it collects and processes large amounts of data, it can become an integral part of a complex system responsible for full service of flying objects. Therefore, essential issue is to ensure cooperation with other IT systems. The software developed under the project meets principal assumptions of SOA, and communication between particular system components is based on services, however ensuring cooperation with other IT systems, especially those that are not based on services, without using the integration platform is difficult. Therefore, it becomes justifiable to use such an integration platform based on BizTalk Server. The integration may involve both cooperation on the level of data integration - a source of the system can be both GRANICE system and any external system - as well as cooperation on the level of process automation. Ensuring cooperation with other systems shall significantly extend the capabilities of the system itself and improve processes associated with maintaining the entire infrastructure including both software, processes and flying objects. Access to the services available on the platform GRANICE in case of the use of integration platform based on Microsoft BizTalk Server is possible through the use of an adapter WCF (Windows Communication Foundation) and appropriate for the database engine adapter (such as MS SQL Adapter), because information collected during the mission are stored in a database. In the future it is planned to create a dedicated adapter BizTalk for communication with the system that shall unify access both to data and services offered within the system. The way to exploit the opportunities offered by the GRANICE system integration with external systems depends largely on the goal that we want to achieve, that is why it is important to be focused on business issues rather than on technical ones. Scenarios for using integration may include a number of areas of cooperation. The most typical one is the scenario in which an external application requests access to data collected during the flight for its use for example to analyze. The GRANICE system data flight parameters are sent in real time (with a slightly delay of not more than 1 second) from an object to authorized monitoring applications, they are also archived in a database. The data format within the system has a specific structure, well-known and acceptable for applications that are a part of it, the format of the data is not known to an external application, which expects data in its format to function properly. It is therefore necessary to convert data from the GRANICE system data format to external applications format. Another requirement posed by the external application is that it only expects certain data, such as object position data only. If you use the BizTalk platform the solution of the problem defined in this way is to prepare a map, in which we may specify what data from our system are to be passed to the external application. An example of an external application may be the reporting application whose task is to prepare reports on the mission performed. It expects data containing information about the mission start time, starting points, the level of fuel, the information about when the characteristics points were achieved and information about the end of the mission. GRANICE system provides the data to integration platform that runs the process responsible for transforming data into a format expected by the reporting application. Processed data are delivered to the reporting application, which generates the required reports. Another example of the use of an integration platform on the level of data integration is a situation in which the mission report is sent by e-mail to defined recipients. The list of recipients may be stored for example in a database or text file. Using available adapters, we can also solve a problem described in such a way. It is possible to publish a report created in SharePoint, because the network adapter is available for this system.

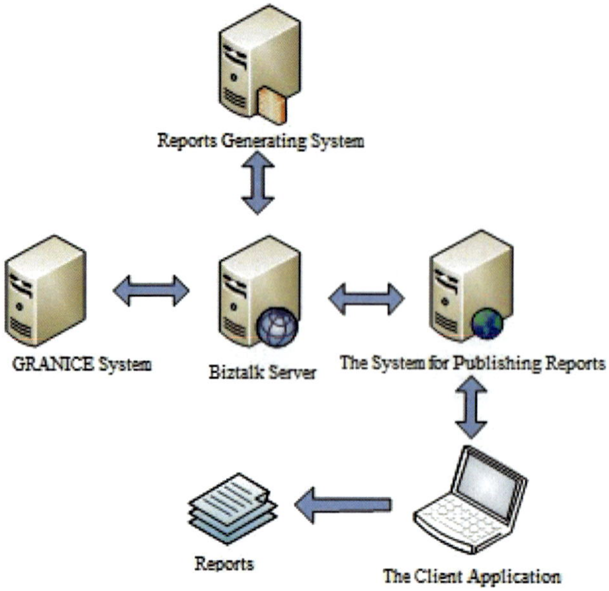

Fig. 5 The integration of the reporting system

Much greater opportunities for integration of applications using BizTalk Server platform are present on the level of process automation and optimization of infrastructure. A typical example of integration on the level of process automation can be a mechanism of building the optimal routes. As a part of the GRANICE system there is available a mechanism responsible for routing on the basis of defined waypoints. The point is defined by geographical coordinates (latitude, longitude and altitude). To plot a route between the points specified at the stage of building the route external services may be used. The role of external service will be appointing (or possible verification) the sets of points defined, the position between the characteristic points defined at the stage of creating the route. Designation of the optimal route may take into account limitations of terrain. An example here might be a situation in which a user specified a route from point A-B, but on the designated route there is a mountain, so the flight of a flying object can be difficult (may require raising the height of flight which could reduce the coverage, etc.). The role of the external software is to analyze all possibilities and choose the best option in order to accomplish a task. This may require adding a few or several new points between those determined by the user. Additionally, it is possible to integrate with weather forecasting system, which will constitute the source of information about atmospheric conditions. Atmospheric conditions have a significant impact on the missions performed by a flying object and can determine their success or failure. Knowing the time and the itinerary of the mission, using a weather forecast system, which is not an integral part of the system GRANICE, you can significantly extend the capabilities of the system. An example scenario is a situation when a given specific purpose of the mission (route, duration and the operations to be performed at each stage) can be supplemented with the data on weather conditions via the integration platform. Identified data may be the basis for deciding whether the mission is to be performed or not. An extensive system of approving plans for the mission may also be built on the base of integration platform.

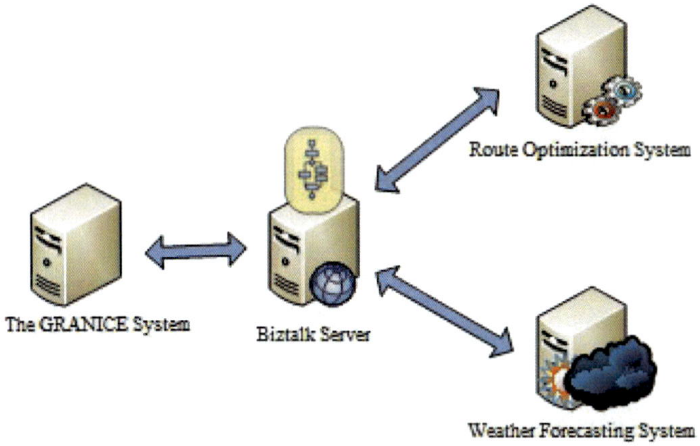

Fig. 6 The integration with routing and weather forecast system

After determining each mission so that it could be performed, there may be the need to obtain approval from a supervisor. Depending on the type of mission (e.g., if the mission itinerary includes a flight in the airspace of another country) it may require the approval of several people. Data concerning mission plan are received from the GRANICE system by the integration platform through a dedicated adapter, and then the workflow process starts, which is responsible for processing the received message and preparing response that includes information whether the mission can be executed or not. The process itself responsible for preparing the answers may be complex and last even several days. Depending on the nature of the mission (duration, monitored area, etc.) request for approval is sent to one or more persons.

This request may take the form of demand transferred to another system or a signed email message. Obtained response can also be in various forms (that is received by the adapter, then using the input stream converted to XML format). Depending on the information that represents the next steps of workflow path are executed. An example here might be a situation in which requires violation of another country space flight and may need the acceptance of minimum two persons entitled to this. Request for approval and details of the mission are sent to three people with such competence. In case of acceptance of two persons to the GRANICE system the feedback is sent that the mission has a positive opinion and can be done, otherwise the system gets information that the mission cannot be executed.

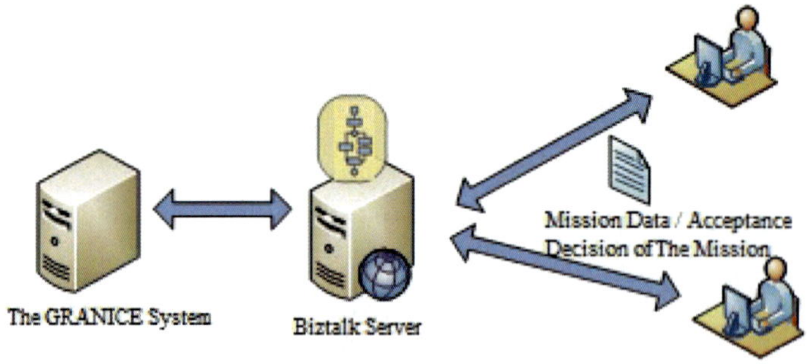

Fig. 7 An example of the integration- the mission acceptance

Executing missions in accordance with specified plan involves sending control commands to the object. These commands include both control of a flying object, as well as the commands responsible for operations carried out by the object (e.g., starting video recording, control of the camera, etc.). At the stage of mission planning there are designated only items necessary to execute the mission and the basic operations defining the tasks performed at a high level of abstraction (eg, start video recording.) For preparation of control commands for a given object we can also use external applications or services. The integration of external applications responsible for the preparation of these orders of GRANICE system can also be executed via the integration platform. Another example of using an integration platform on the level of process automation and optimization of infrastructure may be co-operation with the technical department. Important elements of the GRANICE system are flying objects, which are mechanical devices. Any mechanical device requires maintenance - needed are periodic reviews, repairs of any defects, or even refueling. All operations related to the technical maintenance require documenting in detail (information systems are usually used for maintenance of records). From the perspective of the user of GRANICE system user these operations have no meaning, a user requires only that a given flying object is ready for execution of the mission. Ensuring cooperation between people responsible for carrying out the mission and those responsible for the maintenance of infrastructure related to technical support of flying objects becomes one of the important issues. The mission execution may not be possible if there is no airworthy and prepared flight object. The user expects information about the disposal of the flying object at a time when a mission is to be performed. Technical workers expect information about planned accomplishment of the mission, how many objects will be necessary for its implementation and its duration estimatimation. With this information they can effectively organize their work and plan periodic reviews of equipment.

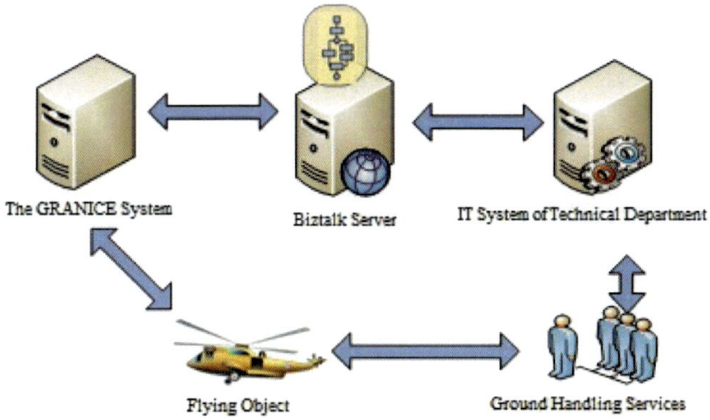

Fig. 8 Integration with the technical department

One of the possible scenarios for the use of an integration platform, may be a situation in which information about authorized missions (along with the planned execution time) go to the technical department information system, which allows the employees of this department preparing the flying object for the flight. The information that the object is ready to execute the mission from the technical department via the integration platform is sent back to the GRANICE system.

5 Summary

Although the GRANICE system is fully functional environment, nevertheless, using the integration platform which is an intermediary in the communication with other systems, we can significantly extend its functionality. Data collected during the flight, then made available through dedicated adapters, and transformed using the maps to the respective formats may be a source of input data for a number of external applications. Their use depends largely on the possibility of applications that use the data. The integration platform based on Microsoft BizTalk Server provides solutions that enable consistent, based on accepted and recognized standards, scalable mechanism for cooperation applications. As a part of the GRANICE system it will allow for creation of integration solutions in a simple way (most of the operations can be performed with the use of graphical editors). The integration platform will become the central element of the system in which there is concentrated information about the processes responsible for the tasks in charge of messaging among different systems. It will allow for creation of more and more sophisticated solutions that meet users' requirements. The cooperation of many different IT systems shall cause that certain processes are carried out more efficiently, and the end user receives valuable information, which is the basis for the reference of a measurable business advantage.

References

1. Banaszyk, P., Fimińska-Banaszyk, R.: Podstawy organizacji i zarzadzania. Wydawnictwo Wyższej Hanzeatyckiej Szkoły Zarzadzania, Słupsk (1997)
2. Brand, P.: Budowa usług z wykorzystaniem JCR (praca magisterska), Akademia Górniczo-Hutnicza im. Stanisława Staszica, Kraków (2010)
3. Brodie, M.: The promise of distributed computing and the challenges of legacy information systems. In: Proceedings of the 10th British National Conference on Databases, Heidelberg, FRG (1992)
4. Goel, A.: Enterprise Integration EAI vs. SOA vs. ESB, http://ggatz.com/ (visited August 16, 2011)
5. ISO/IEC. Open Distributed Processing, Reference model (RM-ODP: Foundations, 1st edn. International Organization for Standardization, International Electronical Commission (1996)
6. Josuttis, N.M.: SOA in Practice. O'Reilly (2006)
7. Lemmens, R.: Semantic interoperability of distributed geo-services, Delft (2006)
8. Pelczarski, T., Widomski, P.: Microsoft Microsoft BizTalk Server 2010 - Integracja, workflow, monitorowanie procesów (2010), http://www.microsoft.com/pl-pl/showcase/details.aspx?uuid=85e4391b-ca20-4bf5-a62e-f0bfe029b12a (visited August 16, 2011)
9. BPEL4WS 1.1 Specification (2005), http://xml.coverpages.org/BPELv11-May052003Final.pdf (visited August 16, 2011)
10. Microsoft BizTalk Server, http://www.microsoft.com/biztalk/en/us/default.aspx (visited August 16, 2011)
11. Microsoft BizTalk Server Esb Toolkit, http://www.microsoft.com/biztalk/en/us/esb-guidance.aspx (visited August 16, 2011)
12. O potrzebie integracji aplikacji - BizTalk Server 2004 - Wprowadzenie (2004), http://www.codeguru.pl/baza-wiedzy/aktualnosc,2109 (visited August 16, 2011)
13. http://www.w3.org/ (visited August 16, 2011)

Tools and Methods Used to Ensure Security of Processing and Storing Information in Databases and IT Systems and Their Impact on System Performance

Łukasz Hoppe and Łukasz Wąsek

Abstract. In the article there were presented technical aspects of special purpose information and classified information protection collected with the help of modern IT&T solutions as well as processed and stored in large data bases. Attention was drawn to formal and legal requirements which in an obligatory manner must be fulfilled by such systems. In the article the issue is discussed of the necessity to obtain system accreditation by the Agency for Internal Security (ABW), which is appropriate in internal security matters of the Republic of Poland. Security measures that should be applied in devices and IT systems are discussed in detail concentrating on physical security measures of IT systems and databases as well as programmary security measures. In the article one could not omit the issue of consequence of a large number of security measures indicating side effects which one has to take into consideration in designing IT solutions serving the security of state.

Keywords: classified information, physical security, software protection, consequences of using powerful protections.

1 Introduction

Dynamic development of IT technologies throughout the past years initiated the era of computerisation of state institutions and offices. The units fulfilling public functions have at their disposal advanced tools and devices which enable more exact and productive execution of entrusted tasks. Obtained information is processed by modern IT systems providing immediate outcomes thanks to which there is significantly shortened reaction time to ocurring events or simply time of execution of the process connected with executed task. This speed of actions becomes especially

Łukasz Hoppe · Łukasz Wąsek
WASKO S.A., Berbeckiego 6, 44-100 Gliwice, Poland
e-mail: {l.hoppe, l.wasek}@wasko.pl

A. Nawrat et al. (Eds.): Advanced Technologies for Intelligent Systems, SCI 440, pp. 211–220.
springerlink.com

important if we talk about units taking care about security of state where only precise and appropriately fast information may constitute efficient protection element. One should, however, remember that the more suport from technology the more data to be transmitted, processed and stored which themselves may be important from the point of view of the state security or may constitute classified information. Initially classified information was processed manually. Produced documents were collected in appropriately prepared rooms, they were produced by a limited numer of people and stored in secret offices. Alongside with development of Information Technology limitation of gathering information to paper documentation ceased to suffice. This need extorted appropriate preparation of IT solutions which currently suport the whole process of special purpose information processing and storing, including classified information. More and more frequently while gathering important data a man is replaced with machines, devices and computers. Some tasks e.g. ones difficult to perform by man are performed by controlled devices, robots or unmanned flying objects. An example may be monitoring and protection of State borders, which because of large scope in the area and difficulty in access to some places is not a task easy for man and additionally forces use of large numer of people i.e. guards. Thanks to application of controlled flying objects patrollling state borders from the air it will be possible to collect and process important data such as pictures or photos needed to patrol guarded territories. An important element is, therefore, appropriate obtaining data collected by unmanned flying objects as well as providing them to appropriate IT systems. Data collected by such devices and systems because of their character must be sent in appropriate way. Communication between the device and system must be appropriately secured in such a way that registered information could be transferred in intact form to IT system where it shall be stored, archived and processed. Takeover or disclosure of registered important data, even classified may entail disclosure of secrets guarded by State Bodies and influence security of the whole State.

2 Classified Information – Legal Considerations

Appropriate protection is as important as the acquisition and processing of information. The Polish law imposes on persons, institutions and devices handling classified information many responsibilities and conditions that must be met to be able to correctly process and protect classified information processed in ICT systems. The most important legislation governing the processing of classified information and ICT security of such data are the following:

- The law dated 5 August 2010 on the protection of classified information,
- Regulation of the Prime Minister of 25 August on the basic requirements for ICT security,
- Recommendations of ABW (Internal Security Agency)(or BSC) for: physical security, ICT security, electromagnetic protection, development of documentation

for the specific requirements of system security, analysis and risk management requirements for software.

The ICT Security is achieved through a set of technical and procedural solutions to mitigate possible threats to an acceptable level. The ICT security concept involves three basic questions:

- What is to be protected? It is the object of protection.
- Against what is it to be protected? It is the identification of hazards.
- How is to be protected? Incorporating the means of protection.

The answer to the first question is contained explicitly in the Law, which contains a definition of protected information that is processed in the ICT systems. This concept has been extended in a document with recommendations on the ICT security of DBTI ABW, according to which also the entire system must be protected, along with devices, in which the classified information and the data itself is processed. Classified information should be protected from unauthorized access, disclosure, phishing, theft, unauthorized modification, falsification and the possible consequences of those actions [1]. Therefore, it seems important to identify hazards and counteract them appropriately. The third issue is related to the technical and procedural compliance with the requirements for protection of classified information, which are imposed on the system or telecommunication network, in which the classified information is processed. According to the applicable law, those requirements are determined by levels: organizational, procedural, technical and personal, to which refer the security service recommendations, related to the protection of classified information processed in the IT systems [1]. All the solutions should be included in the drafting of the document: "plan of protection of classified information" in an organizational unit, as required by the article 18 of the Act on protection of classified information.

The main requirements of security of classified information processing environments include:

- Separation of administrative and security zones, in which the classified information will be processed,
- Preparing a security plan,
- Providing continuous access to critical system components,
- Monitoring access to the premises,
- Electromagnetic protection of network devices,
- Preventing monitoring in the network,
- Encryption of database resources and transmission,
- User authentication and authorization
- Ensuring accountability of user actions.
- Limiting and control of access to data,
- Ensuring reliability of database software,
- Creating backups,
- Providing emergency power supply of key equipment [3].

According to the provisions of the Act, for the protection of classified information in an organizational unit the manager of that unit is responsible, who assigns the protection agent responsible for ensuring compliance with regulations on the protection of classified information. Special responsibility for the ICT systems take the persons acting as the administrator and IT security inspector (Article 64 of the Act). The last link are the persons, end-users of the ICT system, who processes the classified information, with their responsibilities determined by the Act [1]. An IT system used to process the classified information must have an accreditation certificate (Article 60 of the Act). Certification of the systems and networks in the civilian sector, in which the classified information is processed, it dealt with by the Department of ICT Security ABW. The basis to obtain the security accreditation certificate is a positive endorsement by ABW of the detailed documents of the System Safety Requirements and Procedures for the Safe Operation, which shall be prepared by the entity applying for the certificate. Prerequisite for the granting of the certificate is also getting the personnel security clearance and credentials to obtain a positive result following an audit [1].

3 Ensuring Security of Classified Information – Tools and Methods

The issue of managing the security of ICT systems that process classified information is a complex subject, often demanding making strategic decisions. Therefore, before attempting to protect the system, a very thorough risk analysis should be carried out. Risks should be identified, faced by information processed in the system, and then, using an appropriate selection of security measures, effective protection against the identified threats can be achieved. This is a difficult task, often impossible to complete due to the complexity and diversity of issues. It should be kept in mind that the level of system security is not constant, it changes over time, and the choice of appropriate safeguards must be subject to continuous updating and modifications. The issues related to risk analysis in terms of protection of classified information have also been listed in the guidelines of DBTI ABW. The ICT security is the product of the security of infrastructure components. In particular, care must be taken about:

- Physical security,
- Security of transmission,
- Security of devices,
- Security of software,
- Virus protection,
- Authorization and authentication of users.

The ICT Security is ensured by protecting information processed in ICT systems against loss of properties ensuring the security, in particular the loss of: confidentiality, availability and integrity [2]. While deciding on a specific security application in IT systems, one should focus on both, the physical (technical) safeguards of the information systems and databases and software protections.

3.1 Physical Security

The physical protection of systems involves placing a device or a communication data system in the safety zone, and the application of measures to ensure protection against unauthorized access, viewing and eavesdropping. The physical security is implemented by using technical means such as: access control systems, alarm devices or technical means, such as doors, windows, locks that meet the criteria required in the guidelines. The electromagnetic protection of a system consists in preventing the loss of confidentiality and availability of classified information processed in the data communications devices. The loss of availability can be, in particular, the result of disturbing the operation of ICT devices with high power electromagnetic pulses. The electromagnetic protection of a system or ICT network is ensured, in particular, by placing the ICT devices in controlled access zones, which comply with the requirements for electromagnetic attenuation, and by the use of fibre optic cables, instead of traditional copper cables [4]. In order to ensure sufficient confidentiality and integrity of classified information being processed, cryptographic protection is used, in a form of encryptors (to encrypt the transmission over networks), firewalls with options to create VPN tunnels or other devices ensuring hardware completion of VPN tunnels [3]. Ensuring the integrity and high availability of classified information processed in the data communications systems or networks is achieved through redundancy of key system components (servers, databases, application servers), by using backup transmission links, UPS devices and use of power strips preventing from power outages and instability in the power grid. Those elements are crucial for ensuring continuity of the system operation [5]. The definition of availability is one of the basic categories of security, specifying that authorized persons can use it in the required time and place [1]. Provision of access control to premises and equipment communication system is achieved through technical measures in the form of access control systems, magnetic card readers, entrance gates, and electromagnetic yokes.

3.2 Software Protections

The area requiring special security measures is the electronic environment for electronic data processing in a form of an IT system. Protections used in it are crucial to ensure the security of classified information processed. Protection and access control to the IT systems is achieved through the use of mechanisms of the system user roles and permissions, the central administration of user accounts in the domain, the use of PKI with certificates issued on the name of registered system users. The first level of securing the IT systems against unauthorized access is to force a user logs on workstations, using a unique login and password or individual certificate. Since the login and password or certificate is assigned to a particular system user, clear authorization and authentication of the user in the system is ensured. Authentication is the process of verifying the declared identity of a person, device or service involved in the exchange of data. A user logged in the IT system should have access to only those functionalities and data to which he has been granted permission by the system administrator. User permissions should be checked in multiple stages: from the operating system rights, through the powers given in the application, to the database permissions. Currently, the database servers offer the ability to restrict access to both the individual tables, and single database records. Such a restriction of information visibility causes the system to realize the principle of "the need to know." Ensuring accountability is one of the basic security functions of ICT systems. In order to meet the requirements, of user activities in the system must leave a trace. Each operation performed by the user must be registered as evidence. It is recommended to store information about who, when and what was done in the system and to what extent. Accountability is achieved by using dedicated systems or functionalities, or built-in database server mechanisms for registration of activities, such as audit logs, database triggers, journal tables, historical tables that store the operations performed, and tables storing the history of objects change in time. The additional (technical) information gathered allow you recreate the object at any point in time, thus providing information about all actions performed by users on the system. Data recorded in the system should be available as evidence, hence it is important to prepare it appropriately in the databases, for the need of presentation and aggregation of reports. Confidentiality is one of the key safety attributes of classified information that should be especially protected. Appropriate level of confidentiality protection is achieved by using encryption of network connections between the workstations and servers using VPN (SSL connections, access to applications via HTTPS), storing data in the database, on an encrypted partition, or by using encryption of individual database records, protecting access to servers with a user account and password. In the case of exchange of classified information between different security zones, or IT systems, additional security measures should be used in the form of hardware encryptors. In order to ensure data integrity, mechanisms are used in the form of checksums, correcting codes and certificates of registered users. For the security of classified information, they prevent accidental or intentional distortion of the data during read, write, transmission or storage phases with advanced techniques.

To ensure availability of an IT system and data stored in, care must be taken that the servers work in primary-secondary mode or on a cluster basis, and data be properly processed and stored. A very important element are backups of the database made regularly, and application functions to detect and correct errors, in order to minimize their impact on the continuity of the entire ICT system, processing classified information. These days, a great threat to the security of information processed in IT systems is the action of mobile and malicious code, and exploiting loopholes in the systems. Malicious code can be entered into the IT system with the aim to disclose data processed in it, especially the classified data. Protection against malicious and mobile code is ensured by using, inter alia, appropriate anti-virus software. Anti-virus protection should be exercised on all components of the system and the ICT network, and it consists in blocking the possibility of executing a code which poses a danger. Protection of systems and networks is achieved by regular updates of anti-virus software (virus definitions), updates to operating systems and database servers with available patches and service packs, blocking unused ports, analysing outbound and inbound network traffic tracking the CVE (Common Vulnerabilities and Exposures) lists and preventing them, running the software with as low system privileges as possible, using firewall software, disabling unnecessary operating system services, or even separation from the external Internet network. An important element is the separation of the functions performed by the database and application servers on separate machines, so that in case of attack on weak points (gaps) of one of the systems/servers, the attackers could not gain control over the entire machine, in which other servers (e.g. database or application servers) can also be installed [6].

4 Consequences of Using Powerful Protections in Databases and IT Systems

The Act and its implementing regulations require institutions and individuals to introduce a number of necessary safeguards in their ICT systems, which process classified information. Also devices e.g. unmanned flying objects applied for protection of borders must also possess appropriate mechanisms of registration of protection of registered data. They exchange large numbers of data in the form of images registered during their everyday work. Registered photos or other information is sent to IT system by a transmission medium which must be appropriately secured against overhearing, seizure or misuse and this provides the necessity to apply additional safety mechanisms. While discussing tools and methods of protection of important from the point of view of State security information as well as classified information one could not omit the issue of the consequences of their application. Large number of applied protections and mechanisms may significantly influence the manner of operation of such a device and the whole IT system, especially their productivity. The example may be the necessity to assure accountability of all actions in IT system, hence the necessity of registering historical data of all registered and modified data. As one may see in the Fig. 1 alongside with the growth of processed classified information grows the number of historical data necessary to process. Multiplication of the amount of processed and stored information entails the increase in the burdening of IT system as well as the necessity to find appropriate solutions. Therefore a very important element is preparation of appropriate design of the whole system safety already at the stage of designing the IT system.

A certain solution to the issue of accountability and thus the problem of storing and processing a large number of historical data in databases are mechanisms of triggers and journal tables (historical tables). Historical tables correspond with their

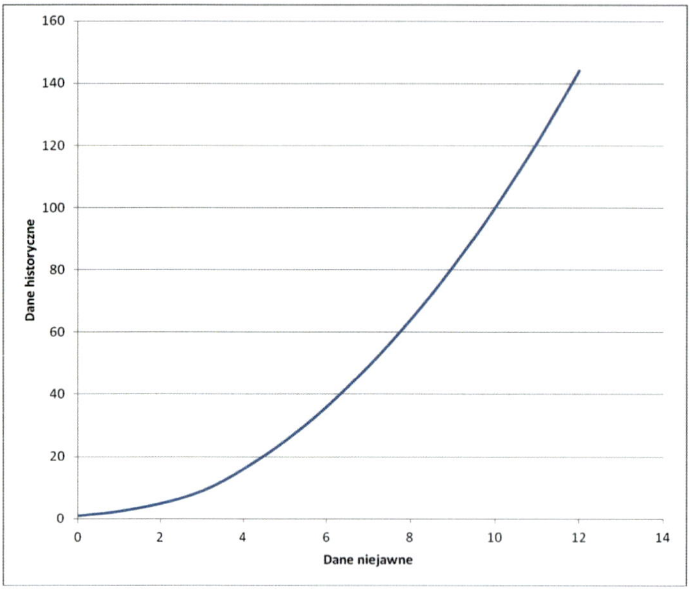

Fig. 1 Multiplication of processed data

structure to source tables but they possess in the structure old and new values of a given record. Historic tables may store both full record irrespectively from the number of altered attributes or only altered values. Depending on the solution it means larger memory engagement or larger time engagement (one needs to longer read changes in a given record). An example scheme of construction of journal (historical) table is contained in the Fig.2.

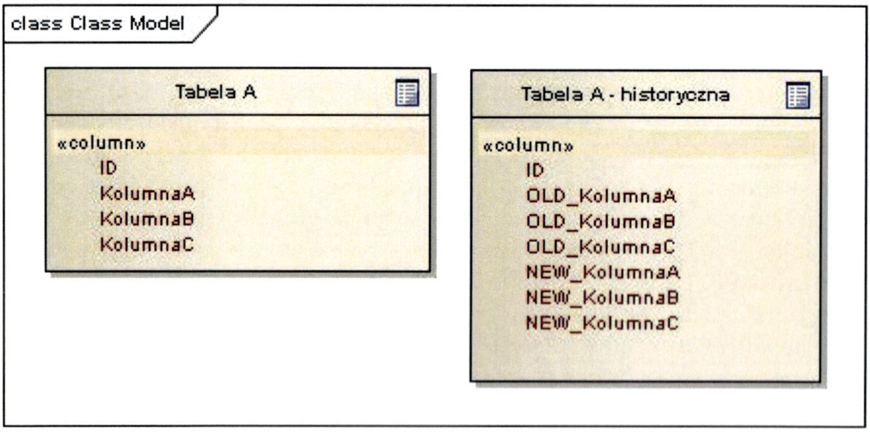

Fig. 2 Structure of source and historical tables

Planning technical solutions designed to ensure the safety of processed and stored classified information is crucial in estimating the cost of the entire solution. If the system performance requirements are not well estimated, as early as at the design stage, operating problems may be encountered at a later time. The result of this can be both, slowing down the system and its complete blockage. A large number of security measures also makes the administrations of its elements difficult. Many of the mechanisms introduced to ICT systems and their interaction in the environment is highly interconnected. Changing one of the parameters of the system can affect others, so it is important to prepare appropriate administrative and operating procedures for the entire system, so that, in the event of an emergency, proper system operation can quickly be restored. A large number of configuration elements, a large quantity of logs collected from the heterogeneous environment makes it difficult to manage it, and it affects the number of people needed to administer the entire system. Such additional information, however, (e.g. server or network device logs) are invaluable when searching for errors, analysing abnormal or suspicious behaviour of the system and data communication networks. The question hast to be therefore asked whether all of those protections must be used in systems that process important for the safety of state or classified information. They cause complications in designing a solution and significantly affect the cost of the entire solution. To answer this question, one should realize how important the presence, in such a class of information systems, seems to be for the critical infrastructure or key equipment for proper operation of the system. In the event of an emergency (i.e. an emergency state, flood, fire) the lack of a reserve processing centre, the lack of redundancy of key components might affect the continuity of the whole system operation; the lack of backups can cause permanent loss of important data constituting state or business secrets, which may jeopardize the safety of the organization or the entire state. Therefore, while designing devices and IT systems used to process classified information, the provisions of the Act, implementing regulations and recommendations of DBTI ABW must be strictly complied with. First, they are vital for the compliance with the requirements of the Polish law, secondly, they ensure correct operation of the system and maintaining appropriate levels of security of classified information processed, sent and stored in different situations.

References

1. Ustawa z dnia 5 sierpnia 2010 r. o ochronie informacji niejawnych. Dziennik Ustaw 2010 Nr 182, poz. 1228 (2010)
2. Rozporządzenie Prezesa Rady Ministrów z dnia 25 sierpnia 2005 r. w sprawie podstawowych wymagań bezpieczeństwa teleinformatycznego. Dziennik Ustaw Nr 171, poz. 1433 (2005)
3. Zalecenia DBTI ABW dla oprogramowania wykorzystywanego w systemach i sieciach teleinformatycznych przetwarzajacych informacje niejawne - wersja 1.2.0 (2008)
4. Szczegółowe zalecenia dotyczace ochrony fizycznej systemów i sieci teleinformatycznych (wersja 1.4), DBTI ABW, Warszawa (2006)

5. Aponicz, L.: Ochrona informacji w przepisach prawnych oraz zasady fizycznej ochrony informacji niejawnych, Tarnobrzeg (2000)
6. Jeda, D., Witkowski, M.: Ochrona informacji niejawnych - nowe rozwiazania. Wydawnictwo Ubezpieczeń, Warszawa (2005)
7. Kazimierski, T.: Planowanie wydajności w środowisku Oracle, Oracle Polska. XII Konferencja PLOUG, Zakopane (2006)
8. Oracle Database New Features Guide 11g, Release 1 (11.1), http://www.oracle.com/pls/db112/homepage

TGIS Application for Monitoring and Control of Unmanned Flying Platforms

Marek Fischer

Abstract. Determining the current position of the flying platform is one of the key elements of the IT system developed for the project GRANICE (BORDERS). Enabling the operator to trace the flight route and current position allows to send a patrol to a precisely defined place and support its operation from the air. The developed system will be equipped with a WWW application for tracing the vehicle position, creating missions by defining checkpoints and specifying tasks for them. The application will support layers and will enable defining alarms and warnings associated with the layers. The layer management module will give the user full control over adding, editing, deleting and displaying layers. The operator controlling the unmanned flying platform should also be able to read basic parameters of the platform's flight. A characteristic feature of the map application is a necessity for working with actual data received from the communication module and historical data stored in a central database. During daily operation the system will display data received directly from the flying platform, whereas data store in the central database will be used to create reports, summaries and restoring historical routes.

Keywords: GIS Web application, internet map module, monitoring of flying objects, user spatial data.

1 Introduction

Defence is the most sensitive, critical and important issue for any country. Logistics and strategy have always posed a serious challenge to every defence operation that has occurred which further requires the knowledge of spatial information i.e. geographical, locational information. Hence, the knowledge of geography is the key

Marek Fischer
WASKO S.A., Berbeckiego 6, 44-100 Gliwice, Poland
e-mail: m.fischer@wasko.pl

A. Nawrat et al. (Eds.): Advanced Technologies for Intelligent Systems, SCI 440, pp. 221–229.
springerlink.com

that benefits defence services. Information like where the enemy is, where ones own assets are, and what lies between the enemy and themselves are very important for any defence strategy. Here comes the very concept of organising information/data as well as their location [2]. Modern Geographical Information Systems is a marriage of advanced information technologies, networking and communications, working on high-performance hardware platforms and widely using lot of peripherals [1]. Determination of the current position of a flying platform is one of the key elements of the IT system which is developed in the project entitled Design and implementation of innovative unmanned mobile platforms for the needs of monitoring State borders. In order to enable the operator to make strategic decisions information about the object location is presented in a graphic form on the digital map of terrain, which is a part of the GIS module. The use of spatial data processing techniques expands the capabilities of data analysis flowing from the unmanned flying objects patrolling the borders.

2 Management of Spatial Data

Designed within the project module of spatial data management will consist of two applications:

- GIS module in a web application that uses map server - access to geographical data collected by the system coming from mobile platforms as well as user-defined system will be possible via a web browser.
- GIS module on a mobile terminal using ArcGIS Runtime library - access to geographical data collected by the system originating from mobile platforms as well as system user-defined will be possible through a dedicated client application.

Due to the diversity of collected spatial data, mainly depending on the source of these data there will be developed database scheme that allows appropriate categorization of these data for rapid analysis. The data describing the location of the flying object at a given moment of time will be treated separately just as the data describing the planned mission of unmanned mobile platform - the user-defined applications. Each type of data will be visualized on a digital map of terrain, as a separate layer with the possibility to define its visibility. Additionally, in order to systematize the work with spatial data in the module there will be separated GIS functional sub-modules that will be operating on the specified data type. Sub-modules are: on-line monitoring of the position of objects, visualization of archival position of the object, defining user layers - the areas, control points, defining of the mission of flying unmanned platforms. Designed within the system the authorisation module will also be implemented in GIS module. Access to spatial data will be defined for specific roles, such as access to the current position of the objects will be limited for operators, other users of the system will be able to analyze historical data, plan missions, define their own layers.

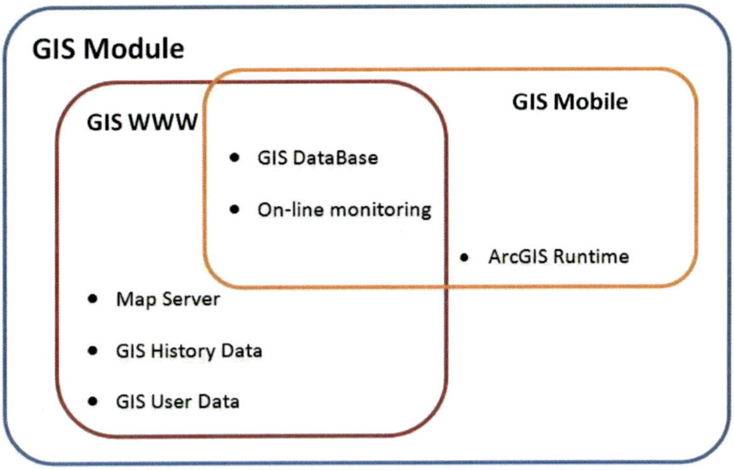

Fig. 1 GIS Module architecture

3 Architecture of Internet Map Server

In general, most Internet map servers adopt a three-tier architecture for the system implementation [4]. The first tier is called "the client tier" which includes the user-side web browser and user-resident Java applets/HTML documents. The client tier is used by the user to make requests and to view maps and remote sensing data. The second tier is the middleware tier that includes the Web Server and the Server Connectors (such as Servlet connectors or ASP connectors) to bridge the communication between clients and the map servers. The third tier is the data storage tier that includes the map server and the database server. The three-tier software architecture of web-based GIS provides customizable functions for different mapping applications and scalable implementation for different hardware [4].

4 GIS Functionalities of the Module

GIS module implemented in the system will have the following functions:

- Operations on a digital map
  - Zooming in - out from a map, a function performed by defining the level of zoom, as well as a second alternative way by bringing a user-selected rectangular area of the map.
  - The measurement of distance - a function that enables to estimate the length of the path designated by the user.
  - Reading the geographical coordinates from the point on the map

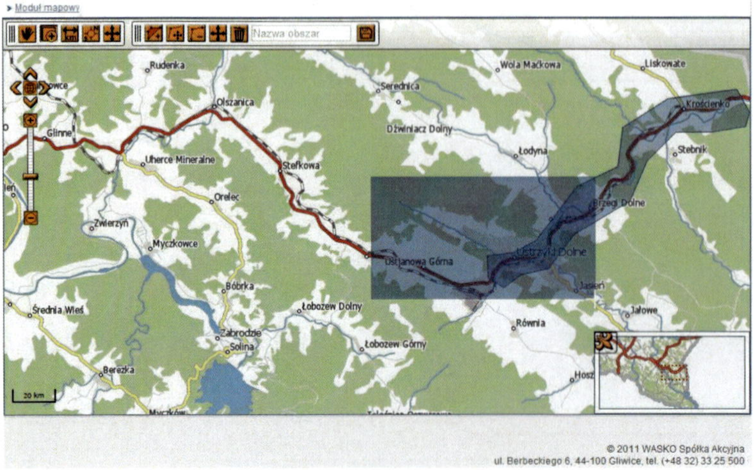

Fig. 2 Map zoom-in (rectangle)

- Entering the data GIS module will include spatial data automatically logged from flying objects, as well as collect information entered manually by the user. In order to allow manual data entry there will be created user module to define layers in which by identifying points on the map a spatial object will be created. It will be possible to define:
  - control points
  - circular areas, a point with the radius
  - areas defined as any polygon

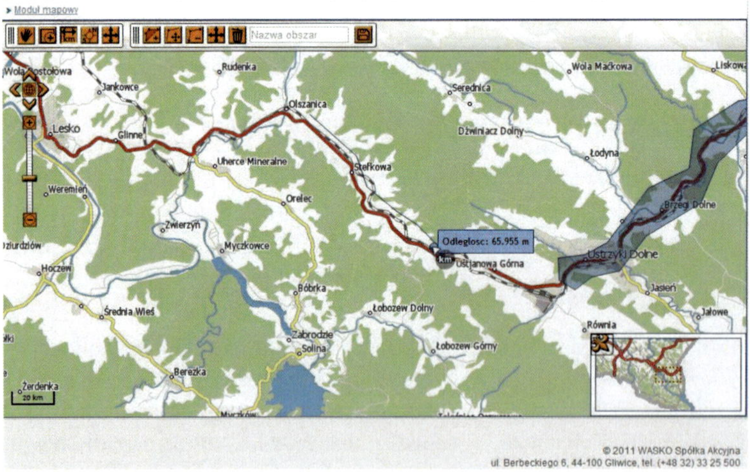

Fig. 3 Path distance

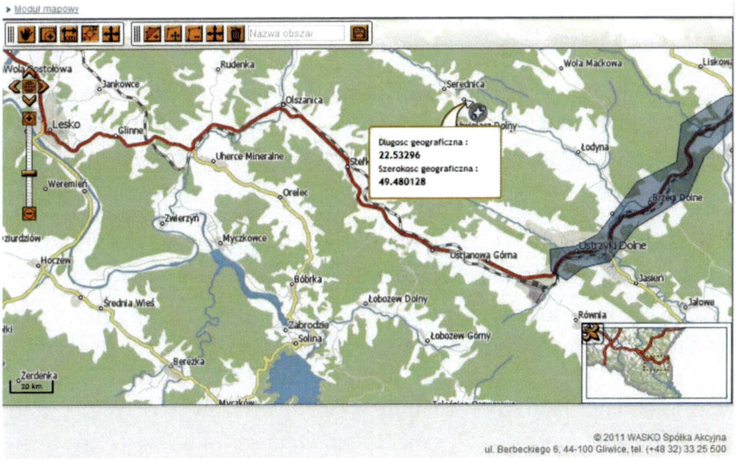

Fig. 4 The geographical coordinates reading

– paths defined as a broken line
– tunnels defined as broken line with the radius

Defining particular elements of the user layer will be implemented by identifying points on a digital map. The module of handling layers will allow the user for full support for adding, editing, deleting, and displaying layers.

- Geocoding
  Used map server provides geocoding function, which allows placing the object on the map knowing only its address - without the knowledge of its geographic coordinates.

Fig. 5 User-defined object on the map

5 The Module of Monitoring Position and Parameters of Mobile On-Line Platform

All the operational activities are a function of location and time. The aim of the decision spectrum involves making available operational resources at the right location at the right time. The decision spectrum comprises of the Observe, Orient, Decide and Act (OODA) cycle and the success of battlefield management lies in reducing the time of our OODA cycle while increasing that of the adversary. An automated tactical information system will help to compress this cycle [3]. An operator handling an unmanned flying platform should also have the possibility to read the basic parameters of the flight platform. Web mapping application will provide these data in real time. In order to provide the support of the operator in on-the -spot decision-making as well as to manage a group of flying objects, the visualization of position and flight parameters on-line module will be created. Data sent to the system from devices will be up-to-date displayed in the window of the mapping module. In addition to the item presented on a digital map of terrain, there will also be visualized such flight parameters as:

- coordinates,
- course in degrees,
- speed,
- velocity relative to air,
- climb / descent rate,
- height,
- value of the throttle,
- engine rotation.

The operator will see information about a selected object as well as will be able to observe the spacing of several selected objects. Such a review of the situation will allow for more complete picture of the situation in the monitored area.

6 Revision of Archival Items and Parameters of Mobile Platform Module

In case of restoration of historical route in addition to retraced course of the route the basic flight parameters will be correlated with the location of an object in space. Additionally, the user will be able to analyze the observed data in the charts available from the mapping module. Gathering functions to analyze geographic information and flight parameters of an object in one place will speed up the time for data analysis.

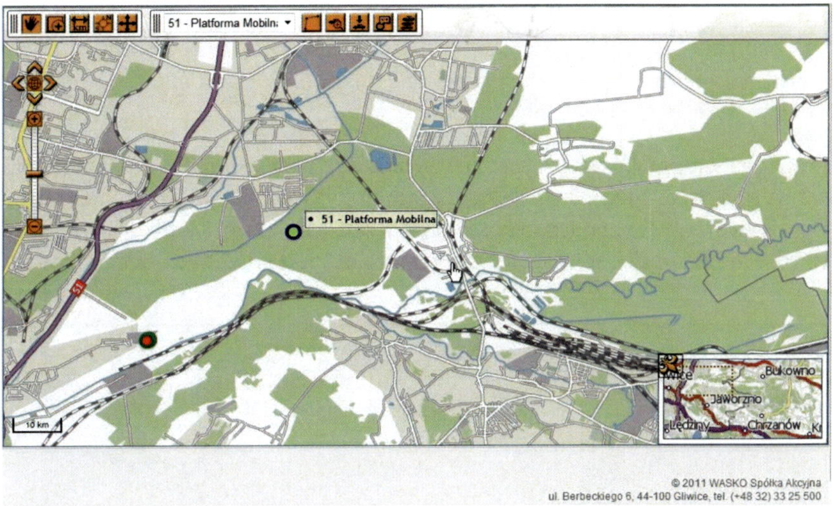

Fig. 6 Mobile platform on-line monitoring

Fig. 7 Visualization of historical data on the map

7 Analysis of Spatial Data, Reports

A characteristic feature of the map application is the necessity to work on up-to-date data obtained from the communication module and archived data stored in a central database. In daily use the system will present the data directly from the flying platform and the data stored in a central database will be used to create reports and summaries and playback of historical routes. The system will allow creating surveys and reports in relation to spatial data. It will be possible, for example, to determine what objects were in the designated area in a given period of time, the frequency with which raised areas are patrolled, etc. According to the report, the result will be presented in tabular form, in the form screenshot with a view of the map, in a graph. Categorization of spatial data and applied algorithms used for spatial data processing allow quick generation of reports' results.

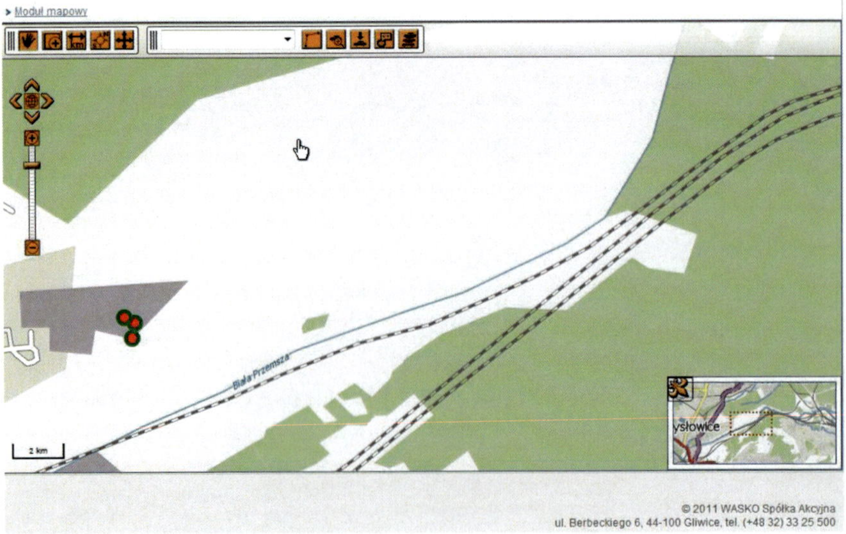

Fig. 8 Map module: report object in area

8 The GIS Application on a Mobile Terminal

Apart from the GIS module in Web applications, monitoring the up-to-date position of the object and the flight parameters will also be possible on a mobile terminal thanks to application created for this purpose. As a part of the module the map module will be created on the basis of ArcGIS Runtime library.

9 Summary

GIS offers a virtually unique ability to aggregate, automate, integrate and analyse geographic data. It is multi-layered. Many layers of information about geographic features can be aggregated on GIS map. Military analysts can use GIS in a number of distinct applications like contingency planning, operational planning, mission briefing, mission rehearsal, deployment and daily operations [2]. Mapping application will be fully integrated with the rest of the system elements so that all functional elements for the user will be presented in a single, consistent interface. Application of GIS techniques for visualization and analysis of spatial data collected in the system will ensure quick and efficient key decisions-making for the protection of the territory of State with flying unmanned mobile platforms.

References

1. Litwin, L., Myrda, G.: Systemy Informacji Geograficznej. Zarządzanie danymi przestrzennymi w GIS, SIP, SIT, LIS, HELION, Gliwice (2005)
2. Singhal, A.: In search of military GIS, `http://www.gisdevelopment.net/application/military/overview/militaryi0001.htm`
3. Srivastava, S.: Requirements of a GIS Based Tactical Information System (2004), `http://www.gisdevelopment.net/application/military/defence/mi04de03.htm`
4. Tsou, M.: Internet GIS: today's realities and a vision of the future, `http://ebookbrowse.com/2002-transactions-gis-tsou-buttenfield-pdf-d61123106`

Vision Object Access and Management with Usage of the In-Memory Databases

Aleksandra Werner, Małgorzata Bach, and Adam Duszeńko

Abstract. Although Poland has been a member of the European Union (EU) for 7 years and internal borders control of the member states have been already lifted (on 21st December 2007 when Poland joined the Schengen Agreement), the safety of a common external borders problem is still actual. Taking into consideration fact that one of longest segments of EU external land border belong to Poland, it is obvious that our country must look for more effective ways of border protect. Our east border is known for being Union's back door used by illegal immigrants that often involve in international terrorism. So, as an area of illicit activities, polish border surveillance system has been developed. It is worth noting that currently Poland has participated in very interesting international research project - TALOS, which is aimed at designing, implementing and testing a prototype of adaptable and transportable border inspection system. But such innovative system, applied in place of conventional systems based on static sensors, still copes with former problem of time-consuming data processing. The high speed data acquisition more than previously involves the development of image processing techniques, so the proposals of metadata separation and proposals of replications usage are described in the paper.

Keywords: vision systems, LOB, in-memory database, replication, federation.

1 Introduction

Nowadays, the automated extraction of information from images is unmatched in its ability of object monitoring, identification or detection of unusual activities. It

Aleksandra Werner · Małgorzata Bach · Adam Duszeńko
Silesian University of Technology, Institute of Computer Science,
Akademicka 16, 44-101 Gliwice, Poland
e-mail: `{Aleksandra.Werner,Malgorzata.Bach}@polsl.pl`,
`Adam.Duszenko@polsl.pl`

A. Nawrat et al. (Eds.): Advanced Technologies for Intelligent Systems, SCI 440, pp. 231–245.
springerlink.com

is widely used in camera inspections applied for example to medical diagnosis, industrial processes, road toll systems or - exactly - border surveillance system. It is noticeable that the spectrum of vision systems application is really very wide, but despite the differences resulting from various fields of their usage, all of them have at least one point in common. All cope with solving the same, image processing, problem. In order to provide the recipients with quick and precise image interpretations concurrently with saving the valuable machine time there is the necessity of improving the image processing solutions. It is worth noting that detection based on relatively simple and fast computations is sometimes used for finding smaller regions (areas) of interesting image data. These areas are further analyzed by more computationally demanding techniques to produce a correct interpretation. Because in one picture a several regions can be separated, desired decrease of a digital image processing time can be achieved, inter alia, by the use of alternative data organization (e.g. in-memory database or column-oriented database).

2 Problem Description

In the area of computer vision, the images are represented by a binary disk files. In monitoring systems, such objects are usually stored in a database - very often as a BLOB fields, corresponding to the images. This approach is conducive to further image processing and guarantees the possibility of executing different types of queries, according to multiply search criteria. Taking into account, monitoring data are usually analyzed by various departments (police, security staff, witnesses to an accident, etc.) that are interested in various details of analyzing objects, this kind of pictures storage seemed to be the best solution. Acquisition the images in machine vision manner, is associated with the simultaneous storage of both the images themselves, as well as many additional data relating to a vision device. This data includes camera intrinsic and extrinsic parameters - such as, for example: focal length, resolution, the position of the camera center, its rotation, etc. Besides, every image has its own attributes that must be stored in conjunction with BLOB object - size, compression, distance from the center of the image to the center of the field of view of the camera, and so on. Some of these parameters - for example the number of bytes per pixel, determine the amount of data that is later processed and have significant influence upon a processing or query execution time. The sample simplified architecture is shown in Fig. 1.

There is also another problem connected with monitoring system entries. Each entry (picture) must be supplemented by its detailed content, in order to provide the semantic meaning of an image. For example it can be the image category or image description. Thus, as it is seen, the full description of one image can require the tens or hundreds of additional data generation. This - in turn - results in the search space expanding, during the descriptions of objects searching.

The fundamental remark is the fact that the images metadata is rather changeless. Generated once for the image does not change or update, and the basic (and perhaps

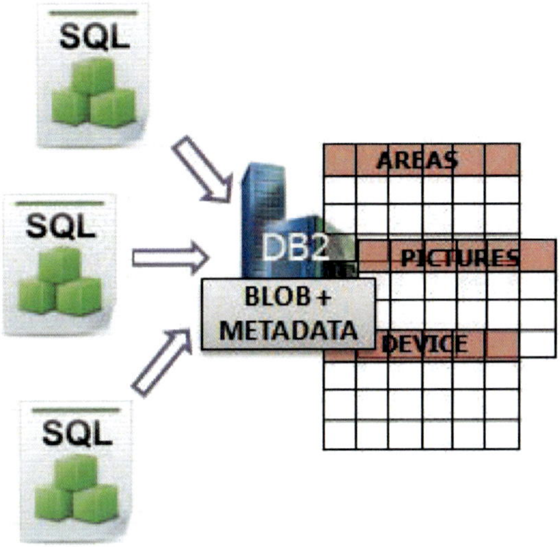

Fig. 1 General idea of discussed problem

even the only) type of performed operation is reading. In order to speed up the search operation, the classical data management systems use buffers located in memory. Access time to cached data reduction, by avoiding the disk operations involvement, is a very important part of query optimization time.

For systems that store a lot of data (e.g. round-the-clock video systems), this approach may be insufficient. Hence, the reasonable solution seems to be using not only memory buffers for data copies caching, but rather the mechanisms of in-memory databases, that entirely reside in memory.

3 Solution Description

Storage of the images and their metadata in the database is well known solution. Since a database systems are provided with the Large Objects (LOB) support, the large amounts of image data have no longer a disastrous impact on the performance of the whole model. It is achieved by the usage of special LOB indicators that point at separately managed storage space. Due to the nature of the monitoring process, the size of this space is still increasing. For that reason, an appropriate organization of the image's metadata, which allows maintaining the required performance growth in spite of space increase, has been suggested.

The general idea of the argued solution is to store images (i.e. BLOB data) and metadata in a relational database (in research it was IBM DB2). The image descriptions (metadata) are - in addition - collected separately. For this purpose, replicas for

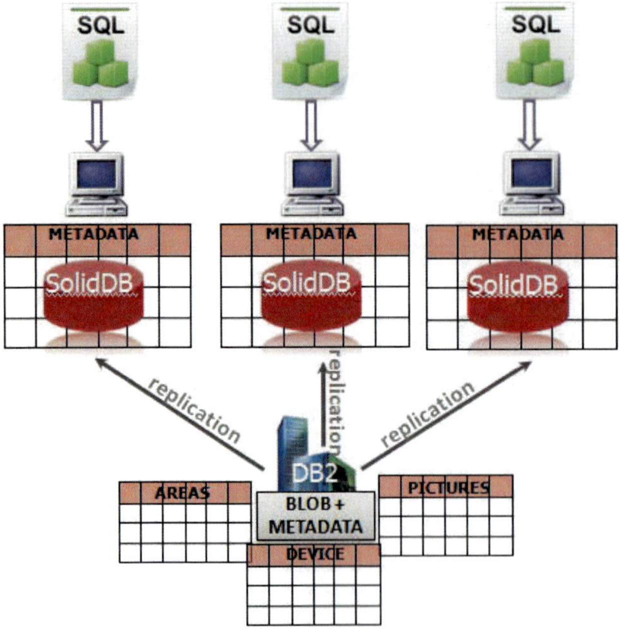

Fig. 2 General idea of proposed solution

metadata stored in a memory-resident databases (IBM solidDB system was used) are created. The visualization of a proposed conception is shown in Fig. 2.

Software that acquires the images saves them in the PICTURES table, but for simplicity of further processing, smaller objects (regions) interesting for observer are extracted from each of the maintained camera shots. The semantic data that refers to these objects are saved in AREAS table. There exists one-to-many relationship between pictures and metadata of their constitutive objects because every extracted region must correspond only to exactly one BLOB data.

```
CREATE TABLE PICTURES (ID INTEGER NOT NULL PRIMARY KEY,
PICTURE BLOB (5M) LOGGED NOT COMPACT, WIDTH INTEGER, HEIGHT
INTEGER, DPI INTEGER, AVG_COL INTEGER, POSX REAL, POSY REAL,
DIRHORIZ REAL, DIRVERT REAL, ALTITUDE REAL, DESCRIPTION
VARCHAR(100), DTREG DATE, DEVICEID INTEGER),

CREATE TABLE AREAS (AREAID INTEGER NOT NULL PRIMARY KEY,
COORDX_LT INTEGER, COORDY_LT INTEGER, COORDX_RB INTEGER,
COORDY_RB INTEGER, CATEGORY VARCHAR(50), ID INTEGER),

CREATE TABLE DEVICE (DEVICE_ID INTEGER NOT NULL PRIMARY
KEY, ANGLE REAL, DISTANCE REAL)
```

The structure of the PICTURES table that stores the images recorded by a measuring device (i.e. camera or unmanned mobile ground unit) is shown in CREATE TABLE statements. The attributes of this table are mostly the image parameters - e.g. image resolution (DPI), width, height of image, etc, but there are also another data - such as the location of the specific object - it means GPS data (PosX, PosY) and extra information about the direction of the device identified by DeviceId attribute, at the moment of taking the photo (DirHoriz, DirVert). Sample shot with the camera directed downwards at an angle of 32° is presented in Fig. 3. The angle of camera's view is stored in the DEVICE table.

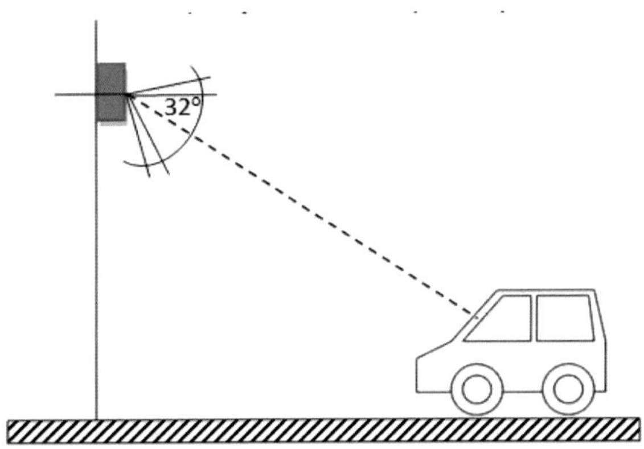

Fig. 3 Sample location of the camera recording the object

The PICTURES table is a type of transaction time table that meets event temporal model, because it is append only (no rows of BLOB data are removed). In such model, the database is seen as a sequence of events - so, here, the history of the monitoring object is determined by a set of consecutive events (for example: object appeared, moved from left to right and disappeared). Apart from BLOBs, a lot of metadata is loaded to this table. They are usually static, because they are known at the time of data insertion[1] or result from the features of vision hardware[2]. Due to this fact, the PICTURES table was at first stored in the DB2 system and then replicated, in order to make the in-memory database performance comparisons. The AREAS table, related to the PICTURES table by an ID attribute, was similarly stored at first in the DB2 system and then replicated in the solidDB database, but the schema of its replicas differed from the source table. In solidDB, static data of AREAS table - area identifier and the coordinates of left-top and right-bottom area corners[3] (Area_Id, CoordX_LT, CoordY_LT, CoordX_RB, CoordY_RB) was completed by selected aggregates (here: AREA_OF_RECT and PAR_OF_RECT).

[1] An example might be the attribute DtReg - the timestamp the photo was taken.

[2] For example DPI or Altitude parameters, where the last one indicates the altitude of a camera at a given moment.

[3] It was assumed, that the shape of an area is the rectangle.

For the research it was assumed, the camera is on Polish territory, so all measured coordinates are within the limits of about 48°-55° (latitude) and about 14°-25° (longitude).

Example 1

The typical query that searches described data can concern finding the ID of all the shots from the cameras that recorded the traced (e.g. suspicious) object. In this case, the query sent to the database contains compound conditions, relate (inter alia) to the direction of photo taking (DirHoriz and DirVert attributes) and the coordinates of both - object (attributes: CoordX LT CoordY LT, etc.) and camera (PosX, PosY). These values are analyzed taking into consideration the angle of camera's view, in order to determine whether the specific object was within camera shot.

Let's assume that (Fig. 4):

1. The point of camera position is labeled with coordinates (PosX, PosY).
2. The point of monitored object (red-colored point) is labeled with coordinates (51, Y_{szuk}).
3. The angle of camera's view equals 70° (shadowed area), so the points placed in the edge of the area are labeled with coordinates (51, Y_1) and (51, Y_2), where the Y_1, Y_2 intercepts are

$$Y_1 = tg(\alpha - 35) * (51 - PosX) \ ,$$
$$Y_2 = tg(\alpha + 35) * (51 - PosX) \ . \quad (1)$$

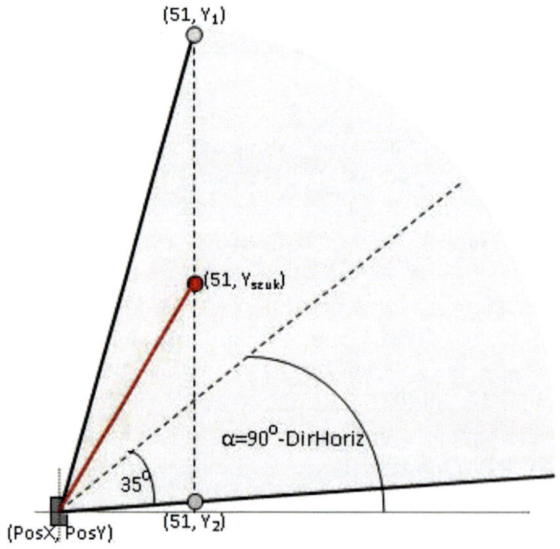

Fig. 4 The interactions of the camera and monitored object

Hence, the following conditions must be met:

1. $Y_1 < Y_{szuk} < Y_2$.
2. The distance (red-coloured segment) between the points (PosX, PosY) and (51, Y_{szuk}), i.e.

$$\sqrt{((51 - PosX) * 72)^2 + ((Y_{szuk} - PosY) * 111)^2}, \tag{2}$$

should be shorter than camera range [4].

Now, the traced object can be found by performing the following SQL query:

```
SELECT COUNT(A.ID)
FROM PICTURES P, AREAS A, DEVICE D
WHERE A.ID=P.ID
AND P.DEVICEID=D.DEVICE_ID
AND
1<SIN(90-DIRHORIZ-ANGLE/2)*(51-POSX)/(COS(90-DIRHORIZ-ANGLE/2))
AND
50>(51-POSX)*SIN(90-DIRHORIZ+ANGLE/2)/(COS(90-DIRHORIZ+ANGLE/2))
AND SQRT(DISTANCE)<SQRT(POWER(((51-POSX)*72),2)+
    POWER(((22-POSY)*111),2))
AND DTREG >'2011-01-01';
```

The search above has very important feature - it can not be supported by a statically defined index. It is because, the query context makes the indexing structure is depending on search parameters. It results in impossibility to define a universal indexing structure that is suitable for starting point of any search. It is a real disadvantage, due to the necessity of sequential search of the entire search space for obtaining the interesting results. In this situation, the required query efficiency seemed to be found in the optimizing the mechanisms that verify a single set of metadata or in replicating a metadata and executing queries in multiply environments.

Proposed solution was validated in research by replicating all, PICTURES, AREAS and DEVICE, tables. Although images are stored in source Pictures table, the in-memory "local" copies of this table store all data except BLOBs. As far as second table, Areas, is concerned, its memory structure is completed by on-line computed area of identified territory.

The existence of several copies of replicated tables is caused by the necessity of simultaneous access to the needed data. It is worth emphasizing that a concurrent access to an image table, resulting from the wide group of monitoring data recipients (police, staff of urban monitoring department, etc.), always causes query performance decline. As opposed to centralized queries processing, the independent executing on local table copies decreases query processing cost. Optimization of the calculation of query conditions can be achieved in different ways. For example by

---

[4] The values 111 and 72 are the lengths (in kilometers) of 1°- respectively - on the meridian and in latitude.

the usage of deterministic user defined functions. The function with specified DETERMINISTIC option always returns the same results for given argument values. Taking into consideration typical selection conditions of analyzed sample queries it was noticed that computing a distance between a device and a subject is mostly calculating. So, the function as follows:

```
CREATE OR REPLACE PROCEDURE CAM_FLD_VIEW
(IN PosX REAL, IN DIRHORIZ REAL, IN ANGLE REAL, IN SIGN CHAR(1), OUT
V_S REAL)
DETERMINISTIC
LANGUAGE SQL
BEGIN
IF SIGN = '+' THEN
SET V_S = SIN(90-DIRHORIZ+ANGLE/2)/(COS(90-DIRHORIZ+ANGLE/2));
ELSE
SET V_S = SIN(90-DIRHORIZ-ANGLE/2)/(COS(90-DIRHORIZ-ANGLE/2));
END IF;
END@
```

can be defined and then used in a sample query:

```
SELECT COUNT(A.ID)
FROM PICTURES P, AREAS A, DEVICE D
WHERE A.ID=P.ID AND P.DEVICEID=D.DEVICE_ID
AND 1< CAM_FLD_VIEW(POSX, DIRHORIZ, ANGLE, '-')
AND 50>(51-POSX)* CAM_FLD_VIEW(POSX, DIRHORIZ,ANGLE, '+')
AND
SQRT(DISTANCE)<SQRT(POWER(((51-POSX)*72),2)+POWER(((22-POSY)*111),2))
AND DTREG>'2011-01-01';
```

As far as replication is concerned, it can be provided by the usage of built-in database mechanism that propagate the changes of metadata tables to a specially defined databases.

In the proposed solution one master DB2 ESE server and many solidDB servers, maintaining slave copies of source database is used - i.e. replication SQL. Chosen (AREAS, PICTURES, DEVICE) database tables writes are captured by especially configured master database server (SAMPLE) and then replicated by the slave (target) database servers (SAMPLE, TOOLSDB and A2) by the usage of apply program (Fig. 5).

Before data can be replicated from the replication source, the replication source must be associated with the target(s) to which the changes should be replicated. This information is defined using subscription sets and subscription-set members (Fig. 6).

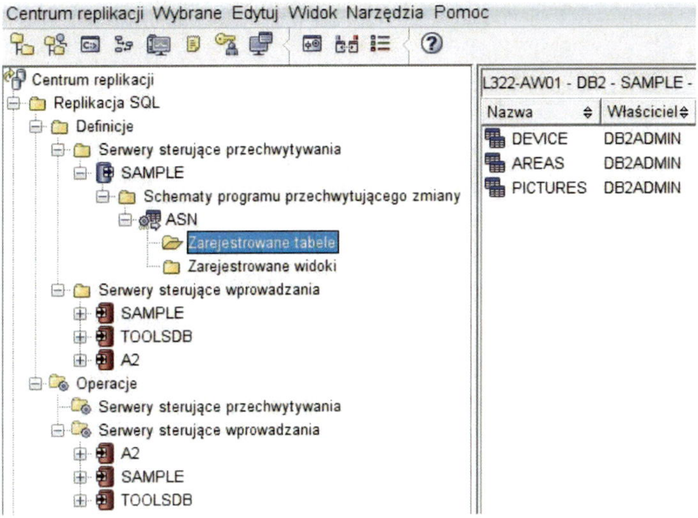

Fig. 5 Replication Center window with defined capture and apply servers

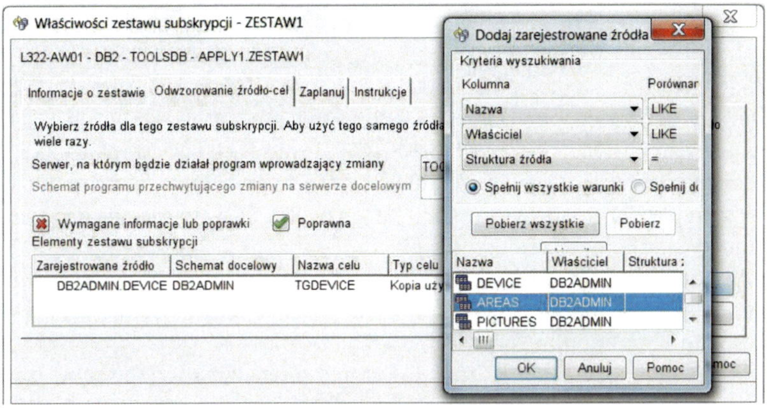

Fig. 6 Defining the subscriptions in Replication Center

In order to build replication for other than DB2 database, the federation infrastructure must be provided. Owing to federation, different relational or non-relational objects (e.g. tables, views, XML files) - even in other than IBM databases - can be accessed as if they are in one DB2 database. For this reason, with the object of research experiments, the DB2 federation with solidDB databases was configured (Fig. 7). Federation's nicknames (e.g. SOLID_PICTURES or SOLID_AREA) allow using DB2's built-in SQL Replication to replicate from source DB2 database to solidDB.

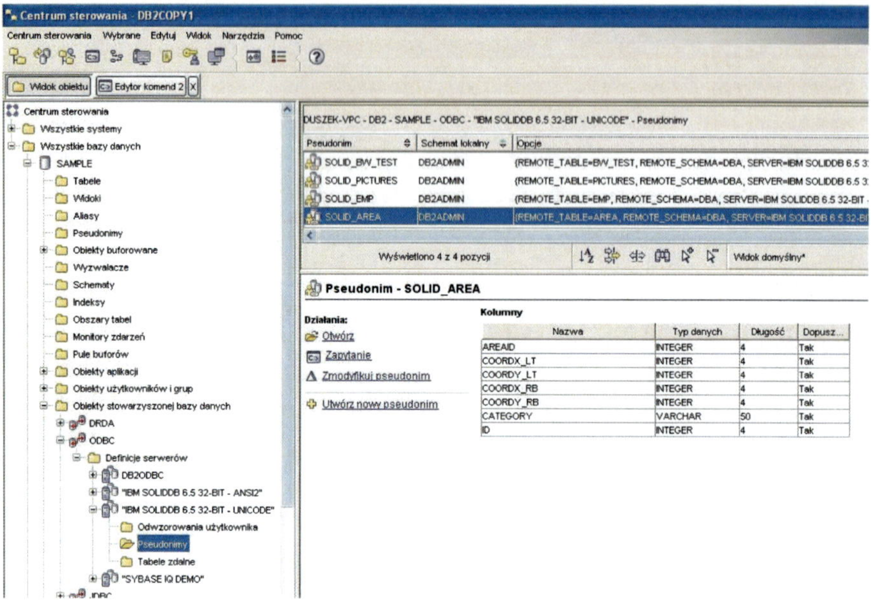

Fig. 7 Federation with defined SolidDB wrappers and in-memory tables nicknames

So, once the solidDB table nickname was created, the replication was tried (DB2 database naming SAMPLE played a role of both, capture and apply, servers).

In discussed research, DB2 federation server was also used to provide a specially developed mechanism propagating the metadata changes to indicated databases. The conception was prepared because of the fact, built-in DB2 SQL replication is relatively - and unnecessarily in described solution - convoluted.

The most obvious way to replicate changes in one table to another is using triggers. This is good solution, but as we found it is not possible in federated databases. Using triggers with federation is very restrictive and cannot be used across tables from physically different databases. Also using stored procedures has the same difficulties. The only way to replicate data changes from master table to federated databases is using external programs fired by triggers. This can be the universal solution even for not federated databases. Putting the connection information into to external programs we can resign from federation. Federation gives us the single point of management of distributed database environment but it is not necessary.

4 Tests

The test environment was configured by the usage of:

- 2,67 GHz computers, with 2 Intel® Xeon® CPU with 24 cores and 12GB of RAM,

- 2.8 GHz PC computers, with 8GB of RAM, Intel Centrino 2 V-Pro processor and SATA 320 GB hard disk,
- 2GHz PC computers with 4GB of memory and Intel Core2 Duo CPU.

In replication tests, source database was stored in the most powerful computers, while the rest hardware served target databases.

For research purposes, different types of SQL queries were executed. Over a dozen queries were analyzed during the tests, but only ten from among them were separated as typical queries displaying monitoring data. Finally, queries with/without tables joining, with/without aggregate functions, with GROUP BY and HAVING clauses, etc., were performed in discussed research.

The number of table's rows varied from 500 000 up to 3 000 000. The size of DEVICE table was negligibly small in comparison with the size of other database tables (remained at the level of a few dozen rows).

In the first place, the execution times of queries performed on master server (with DB2) and on target servers (with solidDB) were collected. To better visualize the differences in speed of query execution in solidDB and DB2 database systems, the logarithm scale on Y axis was used, as it is shown in Fig. 8.

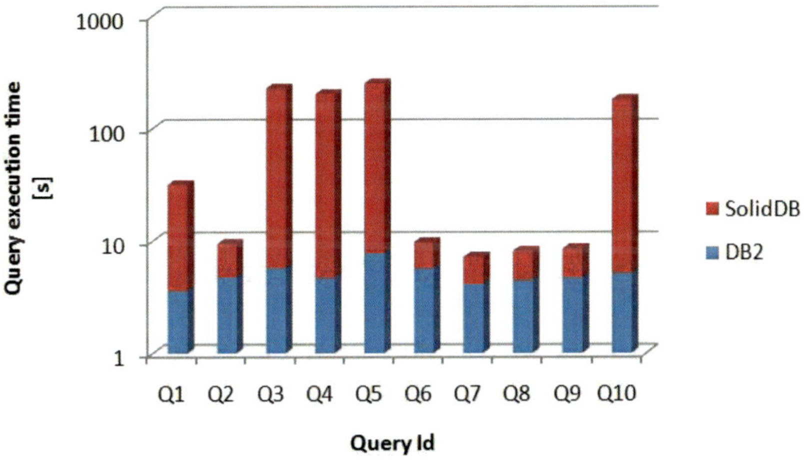

Fig. 8 The cumulative execution time of sample query

Obtained results were unsatisfactory. Only half of the queries were executed faster in a solidDB than in a DB2 database. In some cases, a query execution time 30 times exceeds time measured on DB2 server. Fig. 9 shows how many times slower the chosen queries were performed in solidDB than in classic relational database. Excluding the specific nature of monitoring data, in-memory technique seemed to be unfinished.

With regard to mentioned inefficiency, particular analysis were made with reference to executed queries. It was stated that the most effective queries - identified

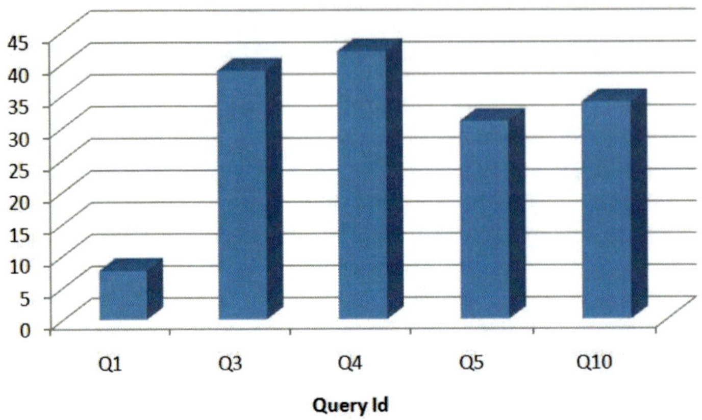

Fig. 9 How many times slower query was performed in solidDB than in DB2

by Q6, Q7, Q8 and Q9 - had the multiple selection operations. For example they included conditions that referred to the picture category or/and to the date of picture taking. On the other hand, queries Q3-Q5 included "heavy" GROUP BY and HAVING clauses.

These findings confirmed our earlier research, relating to the conceptions of in-memory databases, where not only solidDB, but also TimesTen solution was verified. The conclusion of our earlier tests was, queries with low selectivity or for example with NOT EXISTS phrase are less optimally performed in in-memory database than in classical one.

Taking no notice of possible solidDB deficiencies and the necessity of "heavy" queries elimination, in current research the attention is paid to another aspect that can cause results improvement.

Usually, tests are made in artificial – it means: with the usage of university resources - environment. In described studies the situation was similar. The real data but university hardware was used. Besides, only single analyzed query was executed at the time. Thus, it was decided to simulate the real world monitoring environment, where a lot of searches are made at the same time. The number of queries that were executed simultaneously didn't exceeded 10. The effect of query time execution growth is shown in Fig. 10. In most cases, the measured time was about twice longer, but for individual queries (Q2, Q3 and Q9) time was lengthened even up to seven times in relation to time measured for a single query executed on the server. In turn, for query Q8, the unusually time shortening (from 8 seconds to about 7,1) was ascertained. It must be underlined that "single query execution" means, query was executed in environment presented in Fig.2. while "multiple simultaneous queries" – in environment presented in Fig. 1. The chart presents results only for DB2 database, because in solidDB local replicas the indicated problem doesn't exist (probably search only by a single user will be made).

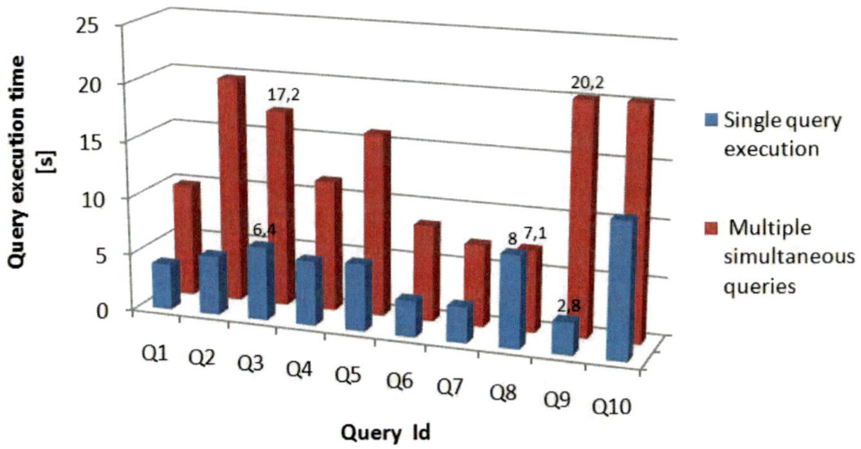

Fig. 10 Test of query scalability

In conclusion, as it was proved, the differences in execution time of queries, performed at solidDB and DB2 servers, are in the real environment lower. This situation is in behalf of in-memory solution benefit. The significant search time extension, observed for some queries executed in solidDB environment, visibly decreases with reference to the time measured for the same queries in DB2 server. Hence, typical searches made in replication environment that are directed to local replicas are in most cases more cost-effective than querying centralized DB2 database.

5 Summary

The relational representation of vision data acquired by monitoring systems consists of several attributes, but relatively small subset of them, is used in a typical query. The performance increase was observed in the variety types of database queries, but acceleration (in some cases) was not satisfactory. The reason may result from the nature of executed queries (a lot of heavy calculations in a query) and from the incremental nature of searched tables (numerous rows). It has been found that up to 100 frames are recorded by one monitoring device in 1 second. This brings 8 640 000 frames (rows) per 1 day. Thus, proposed solution is based on idea of replicating image metadata to the local servers, where in-memory solidDB databases become containers for them. In this architecture, BLOB data is stored only in master server with DB2 database. Hence database reads are divided among all of the database servers, which results in a large performance advantage due to search sharing. The usage of in-memory databases allowed for additional time-saving during searches.

Due to the numerous metadata columns in the analyzed tables, the influence research of data organization to the performance of tables processing, seemed to be

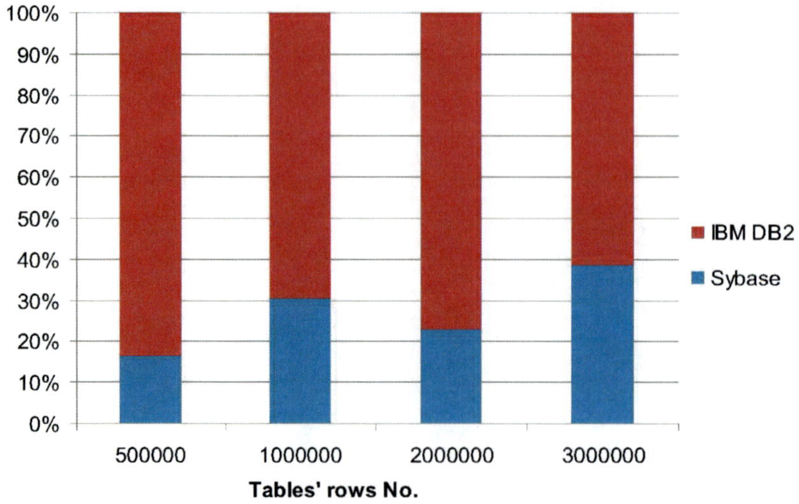

Fig. 11 The cumulative execution time of sample query Q1

reasonable. So, column-oriented architecture with Sybase IQ database was configured. The preliminary tests confirmed the rightness of such approach, as it can be observed in Fig. 11. Results of started tests have indicated that column-oriented data storage significantly speeds up the search operations, but research in this topic is still in progress.

References

1. Chong, R., Liu, C., Qi, S., Snow, D.: Understanding DB2, Learning Visually with Examples. PWN, Warszawa (2006)
2. DB2 9.7 Discovery Kit. Software Press (2009)
3. DB2 9.7 Software Developers Journal Extra no 35, Sofware ISSN: 1734-7661, Warszawa (2009)
4. Bach, M., Duszeńko, A., Werner, A.: The analysis of the database operations acceleration possibility with usage of the In-memory database architecture. In: Kozielski, S., Małysiak-Mrozek, B., Kasprowski, P., Mrozek, D. (eds.) Studia Informatica, Gliwice, vol. 31 2B(90), pp. 63–76 (2010)
5. Duszeńko, A., Maćkowski, M., Nurzyńska, K., Skoroniak, K.: Database usage for multimedia metadata management in monitoring systems. In: Kozielski, S., Małysiak-Mrozek, B., Kasprowski, P., Mrozek, D. (eds.) Database New Technologies, pp. 219–228. WKiŁ, Warszawa (2007)
6. Sybase IQ Administration Guide, `http://manuals.sybase.com/onlinebooks/group-iqarc/iqg1121e/iqadmin/Generic_BookText-View/63`
7. TALOS: EU Border Protection System, `http://talos-border.eu`

8. David, T.: Using Data Federation with solidDB (2011), http://www.channeldb2.com/profiles/blogs/using-data-federation-with
9. Vesset, D.: Reorienting the Data Warehouse: Improving the Business Analytics Capabilities with Sybase IQ Columnar Database. IDC (2010), http://www.dnm.ie/documents/whitepapers/Sybase_case_study_Reorienting_the-_Data_Warehouse.pdf
10. Imai, Y., Sugiue, Y., Andatsu, A., Masuda, S.: Development of web-based monitoring system with database and homing facilities. IADIS (2005), http://www.iadis.net/dl/final_uploads/200507C025.pdf
11. Gedikli, S., Bandouch, J., Hoyningen-Huene, N., Kirchlechner, B., Beetz, M.: An Adaptive Vision System for Tracking Soccer Players from Variable Camera Settings. IAS (2007), http://ias.cs.tum.edu/publications/pdf/gedikli07adaptive.pdf
12. Abuzar, A., Whelan, B., Lyle, G.: Guidelines for Metadata and Data Directory. In: GRDC SIP 2009 (2009), http://www.usyd.edu.au/agriculture/acpa/documents/metadata_documentation.pdf

SETh-Link the Distributed Management System for Unmanned Mobile Vehicles

Karol Jędrasiak, Aleksander Nawrat, and Katarzyna Wydmańska

Abstract. This paper presents a novel approach to the problem of managing mobile robots. It is presupposing the existence of the virtual space in which the calculations are made and the real space containing registered mobile robots. Developed and implemented system combines the best features of agent systems and mobile robot management systems based on a central server. Required system computations were distributed on a separate pieces of software responsible for controlling, called controllers, and monitoring, called monitors. A central communication system that allows to log the state of the system and authorization of connections was used. The solution has been tested using real mobile robots and preceded by an analysis of existing solutions.

Keywords: management system, mobile robot, distributed control.

1 Introduction

During recent years technology of design and creation of mobile robots became more and more advanced and complex. Construction, sensors and effectors were all intensively researched. Mobile robots are commonly used in both military [1] and civilian solutions [2]. Unmanned mobile vehicles are designed in order to perform precisely defined set of tasks, therefore their architecture is highly specialized [3],[4]. Capabilities of a single robot are limited by its construction. As for today, tasks defined for multi-robot cooperation use only dedicated software as in [5]. There is a need for easily configurable system for managing unmanned

Karol Jędrasiak · Aleksander Nawrat · Katarzyna Wydmańska
Silesian University of Technology, Institute of Automatic Control,
Akademicka 16, 44-101 Gliwice, Poland
e-mail: {Karol.Jedrasiak,Aleksander.Nawrat}@polsl.pl,
Katarzyna.Wydmanska@polsl.pl

A. Nawrat et al. (Eds.): Advanced Technologies for Intelligent Systems, SCI 440, pp. 247–256.
springerlink.com

mobile vehicles. One of the main requirements for such system is an ability to change configuration of the system without the need to reimplement most of it. Designing a system that would allow to manage mobile robots regardless of their construction and functionalities is a well known problem. However satisfying solutions for commercial applications are still yet to be found.

Research into cooperation of multiple robots and tasks management received significant attention over last years. Three main approaches are suggested in this work [6]: bioinspired systems, organizational approaches and knowledge-based approaches.

Bioinspired systems assume that mobile robots inside the system are homogenous and in large numbers (e.g. swarms). Each of the robots individually posses highly limited capabilities and cognitive skills. However, when grouped in large numbers and interacting as a collective, a group-level intelligence emerges. Each robot tries to find a solution for an individual sub-problem of a group problem. A failure of the robot during performing task is a no harm for the system. Damaged unit is simply replaced by another. Bioinspired systems were researched in [7] and [8].

Organizational approaches are based on the organizational theory for task allocation. It is assumed that robots can be heterogeneous and capable of different functionalities. There are two main approaches: roles-based and market-based. Role-based management system divides the problem that is to be solved into smaller problems assigned to roles. Role is defined as one or more tasks assigned to the group of robots that are best fitted to the tasks. Example systems based on the roles are described in [9] and [10]. Market-based approach solves the task allocation by allowing robots negotiations. Tasks will be performed by the robot with the lowest task cost factor. Example systems: Sold! [11] and system from the work [12].

Knowledge-based approach assumes that decisions are made using information about mobile robots capabilities in the system. Depending of the solution, each robot contains a task suitability matrix [13], precise model of each other robot in the system [14] or task solution schema generated using individual robot schemas [15].

Mentioned solutions are based on assumption that mobile robots are cooperating with each other. Alternative solution is presented in work [16], where mobile robot is not an active participant of the system but rather a simple tool used to perform specific tasks. A consequence is that the robot's reactions should be deterministic and predictable, based on the current robot state.

Most of the robot management solutions are based on the agent system paradigm. The reason is that there is an easily noticeable analogy between a virtual software agent and a mobile robot which is placed in an environment, equipped with sensors and capable of movement on its own. A mobile robot control program is the only missing part of an agent definition presented in [17]. Therefore it is commonly accepted that a mobile robot can be called an agent and a set of mobile robots is an agent system. The agent system, as noticed in work [18], is a subject of the following disadvantages:

- functionality modification is complicated because of its distributed nature,
- distributed task scheduling is complex and requires dedicated protocols and synchronization mechanisms,
- an error of the hardware is not distinguishable from an error of the software,
- high amount of processing power is required from a single mobile robot.

There exists also multi-robot management systems based on central management server [19]. Central solutions allow to optimize navigation algorithms by utilizing total information gathered in the system. The implemented logic can be also easily modified because it is placed in only one place. Most of the computations are performed on the server machine therefore mobile robots do not need a lot of processing power. The most significant disadvantage is that there is a high risk of error. The error of a central system results in stopping all the mobile robots.

2 Physical and Virtual Space

Tasks to be performed by mobile robots are more and more complex, thus significant processing power and coordination between robots in the system is needed. Complex protocols and navigation algorithms should be used. In work [16] it was noticed that it is very difficult or even impossible to store and run the required software using mobile robots processing units only. It is suggested to divide objects in the system into real space containing mobile robots and virtual space which contains software used for control and monitoring (fig. 1.). Such a division allows to perform excessive computing in virtual space and then only send results in communicates to the unmanned objects. This idea is a basis of further deliberations about an optimal architecture for the mobile robots management system.

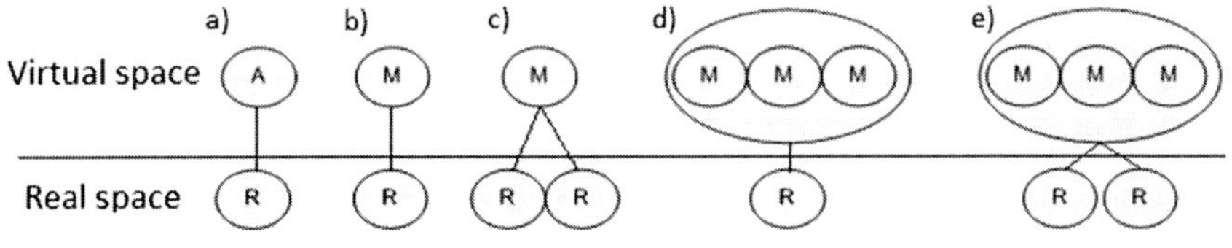

Fig. 1 Virtual and Real space relation types: a, b) one to one, c) one to many, d) many to one, e) many to many. Letter A stand for an agent program, while letter M stands for a monitor program. A controller program is not shown because there is no difference in relation types between monitor and controller.

3 How to Describe the Robot

During designing of the robot management system it is important to take into account a variety of possible mobile robots in the system. There is a need for defining a method for describing robots, which takes into account all mobile robots features currently in the system and all possible extensions. It can be said that mobile robot can be described by (1):

$$r = (S, C), \tag{1}$$

where: S is a state vector and C is a control vector. Such a definition allows an introduction of a meta-description of each of the devices registered in the system. Each device is described by an unique identification number, a name that allows to easily distinguish devices, date of registration in the system and the name of the user who performed device registration. The most important part of the description is a defined state vector. Each element of the vector is an abstract data unit which allows to store number, text and other abstract data units. Such implementation would allow to define hierarchical description of the robots. It is worth to mention that each abstract data unit contains also its own unique id, therefore it is possible to reuse them and combine into more complex structures.

Due to a variety of control methods and possible formats of order messages, it was decided to use analogical solution for control vector as in the state vector. Each mobile robot is defined by meta-description, state vector and control vector which also consist of abstract data units. Sequence, type and values of those units precisely define order message for mobile robot.

4 System Requirements

Designed a hybrid management system for mobile robots combines the best features of agent systems and central control with the main server. Immunity to malfunctions is one of the fundamental characteristics of distributed systems. In case of failure of one of the devices, the only negative effect is a temporary loss of part of functionalities. Replication of a key elements of the system will preserve the full functionality even in case of failure of one of the components. The weak point of the solution that uses a central server for communication is the high possibility of failure. In such case, the whole system stops working. The possible solution is based on using a network of replication servers. Each supporting server is synchronized with the main server and ready to take over its responsibilities in case of failure.

During development of the interface it is required to incorporate the necessary data types that can be transferred. In addition to basic data types, known from programming languages, data from the sensors of mobile robots also needs to be included. Sensor data, depending on the type, can be divided into the following types that need to be considered: scalar values e.g. distance sensor or temperature; the vector values, such as 1D laser scanner or line scan camera; 2D values e.g. laser scanner or video camera; 3D values, such as 3D cameras; depth map recorders; multi-dimensional values e.g. hyperspectral imaging; magnetic resonance imaging. Taking into account the complexity and diversity of data types, it was decided to introduce a communication model based on meta-data messages. Each of the agents of the system software responsible for the inner-system communication should implement the interface of transferring collections of data-change objects. Each of these abstract data change objects includes the data type associated with the logical part of the state vector or control vector of a mobile robot. Such construction could be used in bilateral communication between all software entities in the system.

The presented solution is immune to system failures caused by modifications of the definition of the state vectors or control vectors during system operation. The software responsible for the execution of orders by mobile robots will not fail, but rather will indicate a lack of understanding of the message and notify the request to update the software. It is worth noting that the software update can also be carried out without interrupting system operation.

The communication must be possible between computers equipped with network devices compatible with TCP/IP protocol. However, providing an opportunity to participate in the communication process for other types of devices, such as advanced mobile phones called smart phones, would increase system capabilities. Given the variation in communication technology it seems reasonable to develop an abstract communication interface provided by the virtual proxy object. Such an approach allows easy changes of communication system without any harm for control and navigation algorithms later.

Network communication is fraught with unpredictable risks of breaking the connection. One of the possible solutions to the problem of unstable connection is the mechanism of the session. It is assumed that the agent of a mobile robot sends its unique identifier to the authentication system during connection to the management system of mobile robots. Next it obtains a unique session number. Agent communicates with the system through a proxy object, which is further identified by the system using its own unique session number. After connection interruption longer than time defined as the maximum time for a session, any communication attempts will not be allowed. The server will response to an attempt of communication with an agent via an outdated session with a message encouraging to reconnect and establish a new session.

An important aspect of the management system for mobile robots is to ensure safety. Mobile robots controlled by unauthorized users of the system would cause a danger or a risk of failure. It is necessary to authorize each part of the distributed system that tries to communicate through the system. Authorization via unique id and paired with it password is required to start a session.

Designed distributed management system for mobile robots must meet the following nonfunctional requirements: reliability - resistance to failure; efficiency - the ability to achieve the intended objectives during the predicted time; maintainability – easy modifications and adding new components; portability - the possibility of using the code developed on a different hardware –software platform. Functional requirements are defined as follows: distribution of control and monitoring logic between independent modules and computers; the ability to log the state of the system at any time; adding, modifying, deleting robots in the management system; possible communication between the controllers; establishing a connection and session controlled by the authentication mechanism; the possibility of extending the system with new robots and control and display modules without stopping the system.

5 Agents, Controllers and Monitors

Term agent is used for a piece of software that operates directly on the mobile robot or a computer used to control the mobile robot. Each agent has its own unique GUID identification number and password used to authenticate within server. The main objective of the agent is a translation of the orders received from the server to the communication interface compatible with the mobile robot. As in the agent system, if necessary, it is possible to implement complex control logic for mobile robot in cases of lack of communication with the central server. Therefore, it is possible during an active connection to a central communication server, to obtain the optimal values of control vector by taking into account the state of the entire system. At the same time there is a possibility to implement autonomous navigation based on mobile robot sensors in case of central communication server fault. Sensor data is transmitted to the communication server, which sends it further to the monitors associated with the mobile robot. In order to optimize the operation data is differentially encoded.

Monitor is called a piece of software, which task is to visualize or log the data changes received from the central server. Monitor can be registered to receive data from any number of mobile robots in the management system. During the configuration of the monitor its unique GUID and password is defined. Values of the state vector that will be transmitted to the monitor are also determined during configuration. As only authorized system administrator can perform configuration it can be concluded that the observed values are controlled by the mechanism of permissions.

Controller is referred to a part of a distributed management system for mobile robots which main functionality is the possibility of changing the control vector of chosen mobile robots. As in the case of the monitor it is possible to register controller to any number of unmanned vehicles. During the registration process it is also determined which elements of the mobile robot control vector could be controlled via the controller. Therefore it is possible to define multiple controllers for a single mobile robot.

6 SETh-Link

Distributed management system for mobile robots meets defined in the previous section the functional and nonfunctional requirements was named SETh-Link. The main idea of the system is based on the transmission of messages between distributed parts of the system software through a central communication server. The following types of software involved in the communication: the central server responsible for transmitting messages between the other elements of the system; a piece of software called a monitor used to display the state vectors obtained from the selected mobile robots; a piece of software called controller, used to change the control vector of the selected mobile robots; the software combined with a mobile robot called agent used to transfer messages from and to central communication server.

The system components can be distributed both on a single device (e. g, processes on computing cluster), as well as geographically on different computers over a network.

The main element of the designed system is a central communication server which is the core element of the system. Server is defined as a software which provides the functionality of the logging the state of the system, transferring messages between registered in the system distributed pieces of software and authentication of connections.

Since the failure of a central communication server would crash the entire system, secondary supporting servers are configured and used. Each of the emergency servers is synchronized with the main server in a way that it has a current copy of database of registered mobile robots and applications that are connected. Secondary server queries the main server at regular intervals to confirm its viability. If the main server fails to answer, secondary server is able to take over his duties thanks to the current list of connected clients and their sessions.

Software part of the system is split into: an agent, controller and monitor. Functionalities of the system have been separated into layers of abstraction that communicate with each other through communication interfaces. The system architecture assumes that some layers of the system are optional. Complete solution, incorporating all the elements is composed of six layers. Architecture overview is shown in fig. 2.

The first layer is the physical layer. Mobile robots are registered within it. It is closely linked to the second layer, a layer of agents. Agents layer contains applications of agents that can be run on independent PC computers that communicate with the system or implemented directly in hardware of mobile robots.

The third and the fifth layer include the network devices and networking protocols necessary for the two-way communication between the central server and agent. The presence of these layers in the system is optional because it is easy to imagine a system in which mobile robots are controlled directly from a computer that is running a central communication server.

The fourth layer is the layer of the central communication server, which is the core of a distributed management system for mobile robots SETh-Link. At this level secondary servers and any devices and software directly used by the communications server to implement its functionality, e. g. databases used to log system status are also placed.

The final sixth layer contains applications of distributed management system for mobile robots SETh-Link: agents, monitors and controllers. Each of these applications can be deployed both within the server computer or via Ethernet.

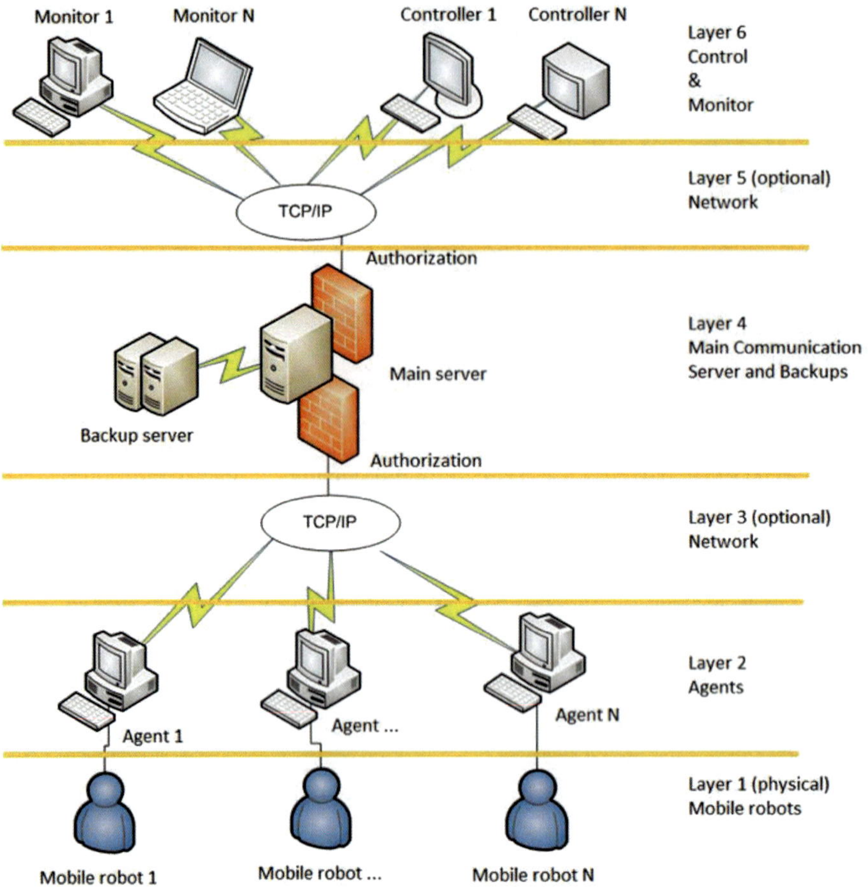

Fig. 2 Architecture of a distributed management system for mobile robots SETh-Link. The logical layers are visible.

7 Tests

Presented the distributed management system for mobile robots SETh-Link has been implemented and tested using the following hardware platform: two mid-range PCs, mobile robot Jaguar [20].

Jaguar mobile robot is a multi-purpose, mobile platform equipped with: a GPS receiver, IMU unit with nine degrees of freedom, the two arms allowing rotation 360 degrees around its axis, video camera.

The tests were conducted to verify whether the proposed solution of the distributed management system for mobile robots SETh-Link is capable to control unmanned mobile vehicles.

8 Conclusions and Future Work

Developed prototype system was called SETH-Link. The system is based on a central communication server and distributed software components. This solution has a logical six-layer architecture.

Except central management there is also the possibility of implementing control algorithms for autonomous mobile robots, if, for some reason, it is impossible to communicate with central server.

Developed and implemented applications of monitor and controller were successfully used to control mobile robot Jaguar and test its capabilities. Developed Seth-Link system could be used to manage both a single robot and a group of robots.

The system could be used to implement security system utilizing teams of robots and sensors to protect the buildings. Each robot could be controlled by algorithms using the state of the surrounding environment as seen by all the robots together. Another application might be managing domestic mobile robots in the absence of the owner of the property. By using an Ethernet network it would be possible preview parameters of the system from anywhere in the world via the Internet. In military applications the system could be used to manage a fleet of unmanned flying objects (called UAVs) supported by unmanned ground vehicles (called UGV). Distributed, flexible system architecture allows the implementation of solutions acting autonomously as those for mobile robots in the management of the property, as well as hybrid solutions where the parameters calculated by the control system can be modified in real time by the system operator.

References

1. Unmanned-Combat-Air-Vehicles, www.deagel.com/Unmanned-Combat-Air-Vehicles (visited April 1, 2011)
2. iRobot Research, www.irobot.com (visited April 1, 2011)
3. Daintith, J.: Mobile robotics. A Dictionary of Computing (2004), http://www.encyclopedia.com/doc/1O11-mobilerobotics.html (visited May 6, 2011)
4. Zhong, C., DeLoach, S.A.: Integrating Performance Factors into an Organization Model for Better Task Allocation in Multiagent Systems. Kansas State University (2010)
5. Collinot, A., Drougoul, A., Benhamou, P.: Agent Oriented Design of a Soccer Robot Team. In: Proceedings of the Second International Conference on Multiagent Systems, Kyoto, Japan, pp. 41–47 (1996)
6. Parker, L.E.: Distributed Intelligence: Overview of the Field and its Application in Multi-Robot Systems. Journal of Physical Agents 2(1), 5–14 (2008)
7. Sun, S.-J., Lee, D.-W., Sim, K.-B.: Artificial Immune-Based Swarm Behaviors of Distributed Autonomous Robotic Systems. In: Proceedings of IEEE International Conference on Robotics and Automation, vol. 4, pp. 3993–3998 (2001)
8. Sugihara, K., Suzuki, I.: Distributed Algorithms for Formation of Geometric Patterns with Many Mobile Robots. Journal of Robotic Systems 13(3), 127–139 (1999)
9. Simmons, R., Singh, S., Hershberger, D., Ramos, J., Smith, T.: First Results in the Coordination of Heterogeneous Robots for Large-Scale Assembly. In: Experimental Robotics VII. LNCIS, vol. 271, pp. 323–332. Springer, Heidelberg (2001)

10. Stone, P., Veloso, M.: Task decomposition, dynamic role assignment, and low bandwidth communication for real-time, strategic teamwork. Artificial Intelligence 110(2), 241–273 (1999)
11. Gerkey, B.P., Matari'c, M.J.: Sold! Auction Methods for Multirobot Coordination. IEEE Transactions on Robotics and Automation 18(5), 758–768 (2002)
12. Sariel, S., Balch, T., Erdogan, N.: Incremental Multi-Robot Task Selection for Resource Constrained and Interrelated Tasks. In: Proceedings of the 2007 IEEE/RSJ International Conference on Intelligent Robots and Systems, pp. 2314–2319 (2007)
13. Fua, C.-H., Ge, S.S.: COBOS: Cooperative Backoff Adaptive Scheme for Multirobot Task Allocation. IEEE Transactions on Robotics 21(6), 1168–1178 (2005)
14. Parker, L.E.: ALLIANCE: An Architecture for Fault Tolerant Multirobot Cooperation. IEEE Transactions on Robotics and Automation 14(2), 220–240 (1998)
15. Tang, F., Parker, L.E.: Distributed multi-robot coalitions through asymtre-d. In: Proceedings of International Conference on Intelligent Robots and Systems, IROS, pp. 2606–2613 (2005)
16. Ambroszkiewicz, S., Centrarowicz, K., Turek, W.: Multi-robot management framework based on the agent dual-space control paradigm. In: Proceedings of the AAAI 2007 Fall Symposium, Arlington, Virginia, USA, pp. 32–37 (2007)
17. Rao, A., Georgeff, M.: BDI-agents: from theory to practice. In: Proceedings of 1st International Conference on Multiagent Systems, San Francisco (1995)
18. Turek, W.: Agentowy system wielomodelowy do zarzadzania grupa robotów mobilnych, Rozprawa doktorska. Wydział Elektrotechniki, Automatyki, Informatyki iElektroniki. AGH, Kraków (2009)
19. Park, J.-H., Song, T.H., Jung, S., Kwon, K.H., Jeon, J.W.: A User Interface Transmission System for Public Robots. In: Lee, S., Choo, H., Ha, S., Shin, I.C. (eds.) APCHI 2008. LNCS, vol. 5068, pp. 141–148. Springer, Heidelberg (2008)
20. Jaguar Mobile Robotic Platform, `http://jaguar.drrobot.com/specification.asp` (visited June 6, 2011)

Delays Models of Measurement and Control Data Transmission Network

Tadeusz Topór-Kamiński, Beata Krupanek, and Jarosław Homa

Abstract. The paper herein aims at analysing one of the parameters characterising data transmission network. This parameter is transmission delay. The first part of the article presents the concept of teletransmission system delay model construction with reference to all sources of delays generated during data transmission between a source and target. This model also includes the network proxy devices which play a significant role in the total amount of delays. In the following part of this paper the analysis of transmission delays has been described with the use of probabilistic model. It has been supported with simulation examples. For this reason, the experimental transmission system was built which was based on ZigBee wireless standard. There were experiments made in different propagation conditions on the basis of which the assumed model has been verified.

Keywords: delay, delays model, network, delays measurement.

1 Introduction

In the times of common usage of measurement and control data transmission networks the crucial problem relates to the determination of quality of services provided by such networks. In the modern automation systems the knowledge of transmission channel quality parameters allows for adequate design of measurement and

Tadeusz Topór-Kamiński · Beata Krupanek
Silesian University of Technology, Institute of Measurement Science, Electronics and Control, Akademicka 10, 44-101 Gliwice, Poland
e-mail: {Tadeusz.Topor-Kaminski,Beata.Krupanek}@polsl.pl

Jarosław Homa
Silesian University of Technology, Institute of Automatic Control, Akademicka 16, 44-101 Gliwice, Poland
e-mail: Jaroslaw.Homa@polsl.pl

A. Nawrat et al. (Eds.): Advanced Technologies for Intelligent Systems, SCI 440, pp. 257–279.
springerlink.com

execution systems. One of the quality parameters of data transmission systems are delays. The long delays resulting from small efficiency of the telecommunication system used in the process of production or control can cause serious technical problems [1] (incorrect measurement results, inaccurate control) which influence the final success of the applied solution. In order to be able to analyse the occurrence connected with delays in telecommunication systems we have to adopt its accurate definition. We have to distinguish two separate types of delays which analysis is slightly different [2][3][4], e.g. we can distinguish one way delay (Fig. 1) and two way delay (Fig. 2)

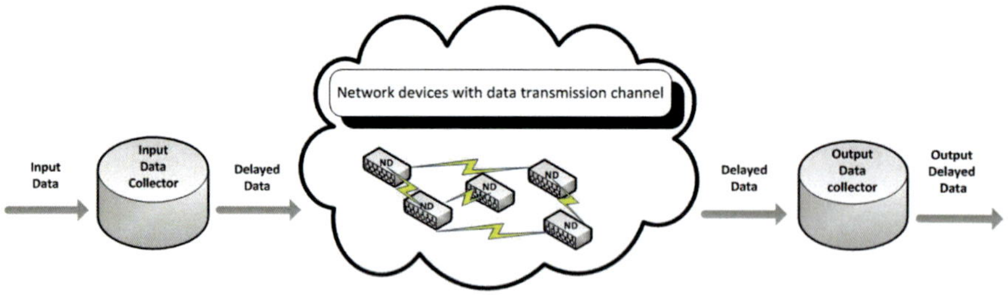

Fig. 1 General model of telecommunication system with one way delay

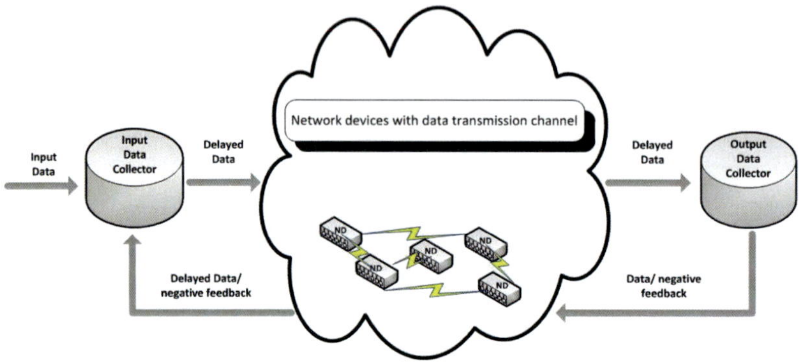

Fig. 2 General model of telecommunication system with two way delay

Each of the described types of delay is important in terms of different kinds of transmitted data. One way delay is connected with transmission of e.g. voice data, video streams, etc. so data which are subject to analysis in the ODC and do not influence further transmission. The two way delay is mainly connected with different types of automation systems - measurement and control system [1]. In such systems there is usually a certain object called the controller, and most often the controlled inert object, between which the exchange of data occurs by means of feedback. This exchange occurs through telecommunication system. However, it should be emphasized here that due to the character of telecommunication system in the described case, the one way delay for the same system is always different from delay in the opposite direction so two way delay is not the doubled value of one way delay [2][4]. Taking the above considerations into account the following definitions has been adopted: one way delay means the time between data input, preparation of data by IDC (Input Data Collector) and the time in which this information will be received and prepared to be used by ODC (output data collector), two way delay means the time between data input, preparation of data by IDC (Input Data Collector) and the time in which this information will be received one more time and prepared to be used by IDC (Input Data Collector) after its previous reception, preparation, processing and retransmission by ODC (output data collector). The delay is influenced by many factors [4], which can be described both in the deterministic, as well as probabilistic way. The deterministic dependencies are mainly used to determine maximum delays in computer networks [5], whereas probabilistic means are used for acquiring the delays distribution occurring as a result of factors described randomly [6,10]. The probabilistic description of delays provides full information on their features [6,10], therefore its introduction into the analysis of network characteristics can have a broader meaning that appliance of deterministic description.

Let us consider the least complicated situation in which IDC and ODC of a computer network communicate with each other without connection by means of any devices. Such situation is illustrated in Figure 3.

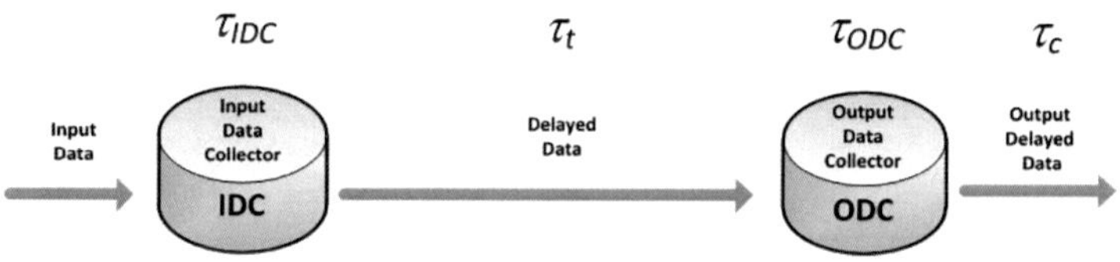

Fig. 3 Source of delays during transmission from IDC to ODC

Let us use the symbol τ_{IDC} to mark time needed for preparation of data and acquiring the access to a carrier (communication medium) in IDC, the symbol τ_t will represent time used for sending the message from IDC to ODC, and t_{ODC} will stand for time of data reception and their preparation in ODC. Let us assume, that the transmission is initiated at the time marked as t_0, and start of data transmission by a network occurs at the moment t_{IDC}. In such situation transmission delay connected with IDC is:

$$\tau_{IDC} = t_{IDC} - t_0 \tag{1}$$

The delay τ_t in message transmission from IDC to ODC is the difference between the time t_t and the moment of transmission initiation t_0,. Such delay, after including dependency (1), equals:

$$\tau_{ODC} = t_t - t_0 = \tau_t + t_{IDC} - t_0 = \tau_t + \tau_{IDC} \tag{2}$$

So it is a sum of two partial delays. The total delay in this case is:

$$\tau_c = t_{ODC} - t_0 = \tau_{ODC} + \tau_t + t_{IDC} - t_0 = \tau_{IDC} + \tau_t + \tau_{IDC} \qquad (3)$$

where t_{ODC} is the time of reception completion and preparation of data in ODC. Therefore, the total delay is the sum of partial delays which source are individual elements participating in data transmission.

2 Transmission Delays Models

On the basis of definitions proposed above, it is possible to build a data transmission delays model Fig.4. Such model includes all sources of delays which may occur in teletransmission system. It is a theoretical model which was established to facilitate analysis of occurrences influencing the delays during data transmission in a computer network [7].

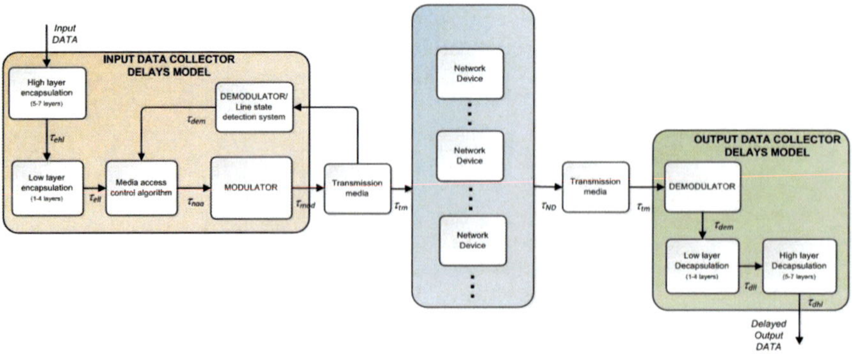

Fig. 4 The data transmission delays model, where: τ_{ehl}, τ_{ell} mean delays introduced by the process of encapsulation of high layers and low layers of OSI model, τ_{naa} means delay generated by media access control algorithm, τ_{mod} means delay generated by the process of modulation, τ_{dem} means delay generated by the process of demodulation, τ_{tm} means delay introduced by transmission media, means delay introduced by a Network Device, τ_{dll},τ_{dhl} respectively mean the delays introduced by the process of decapsulation of low layers and high layers of OSI model

In the model presented in Fig. 4 we can distinguish three basic blocks, i.e.: the delay model IDC (input data collectors), the delay model ND (network device) and the delay model ODC (output data collector). There is an adequate transmission medium between the mentioned blocks, which generates delay characteristic to its properties.

The delay model IDC is characterised by those actions which have to be undertaken in order to prepare and transmit data through transmission medium. The first element of the model called `high layer encapsulation` reflects delays connected with actions represented by the three highest layers of OSI model

[2][3] (application, presentation, session). Those layers are mainly responsible for proper coding, data presentation, as well as for creation and maintenance of session between IDC and ODC. Another element `low layer encapsulation` represents the actions of 1-4 layer of the OSI model (transport, network, data connectors and physical). The transport and network layers are characterised by adequate protocols adopted during transmission but nowadays they have been dominated by the following pairs: TCP/ IP and UDP/IP.

The separation of high and low layers in the model is justified by the fact that the actions of high layers are mainly realized at the level of application software, and the actions of low layers are mostly connected with operational system and hardware systems concerning the access to the medium, e.g. NIC (network interface card) [2][3].

Other elements of the model are connected with media access layer and physical layer of OSI model. The first one called `the media access control algorithm` is the software element strictly cooperating with hardware systems of Modulator and Demodulator. It is usually represented by two algorithms: CSMA/CD (Carrier Sense Multiple Access / with Collision Detection) which is the algorithm used for detection of carrier with multiple access and collision detection, and it is characteristic to the wire Ethernet technology. CSMA/CA (Carrier Sense Multiple Access with Collision Avoidance) which similarly to its precursor is the algorithm used for detection of carrier with multiple access but with collision avoidance is characteristic to wireless Ethernet technology. The system of Modular and Demodulator are usually specialized integrated circuits responsible for the conversion of data into the signal adequate for certain transmission medium, i.e.: electric signal, optical signal or signal in a form of electromagnetic wave.

Taking the delays connected with the elements described above intro consideration, the total delay generated by IDC marked as τ_{IDC} can be presented in a form of a total of delays introduced by each of the elements of the model presented in Fig. 4, consistently with the dependency

$$\tau_{IDC} = \tau_{ehl} + \tau_{ell} + \tau_{naa} + \tau_{dem} + \tau_{mod} \tag{4}$$

The addressee of data in the model proposed in Fig. 4 is the object called ODC (output data collector). It represents those actions which should be conducted in order to recreate data transmitted to its input from transmission media. The first necessary element in the chain of actions of ODC is Demodulator transforming the physical signal into the digital message. Then, there occur two elements introducing delays connected with the process of decapsulation, i.e.: low layer decapsulation and high layer decapsulation. They realize a process that is contrary to the process of encapsulation described above. This process has been divided into two parts connected with high and low levels of OSI model, for the same reasons as in the case of IDC. Taking the above into consideration, the delay introduced by ODC according to the model presented in the figure is represented by the following dependency:

$$\tau_{ODC} = \tau_{dem} + \tau_{dll} + \tau dhl \tag{5}$$

In a special circumstances only adequate transmission media occur between IDC and ODC. However, in reality we deal with a situation where data need to travel a longer distance through different types of transmission media. In such situation, it is necessary to use adequate proxy devices. In the considered situation these are elements marked as network devices. We can distinguish many types of network devices, but we can also list main groups of such devices, i.e.: switches, routers, firewalls or UTM, proxy servers. The role and functions of each group are significantly different. They are based on different layers of OSI model and they perform different functions. However, we can indicate a general structure which reflects the sources of delays for every network device. Fig.5 shows the universal delays model for network devices.

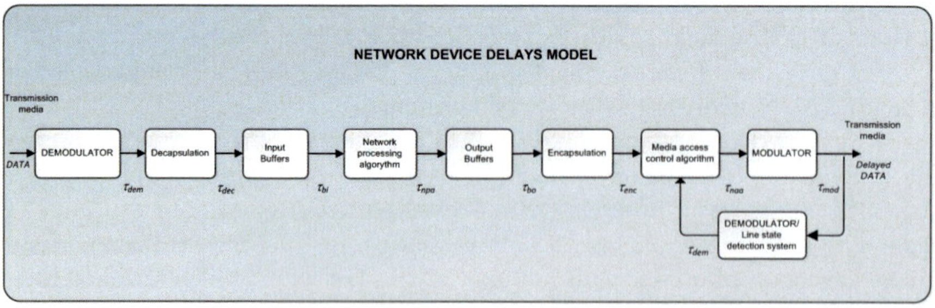

Fig. 5 Network device delay model, where: τ_{dem} means delay generated by the demodulation process, means delay generated by the decapsulation process, τ_{npa} means delay generated by the network processing algorithm, τ_{bi},τ_{bo} mean respectively: delays connected with input buffers and output buffers, τ_{enc} means delay generated by encapsulation process, τ_{naa} means delay generated by media access control algorithm, τ_{mod} means delay generated by modulation process

The model presented in Fig 5. contains all elements which are common for all network devices and are a source of delays in data transmission. The first element of the model is `demodulator` performing the same function as for the ODC. Then the process of decapsulation has to be made, but it will include different layers of the OSI model depending on the type of a device, e.g. for switch these are 1-2, for router these are 1-4, and for UTM device these are 1-7. The role of another element called `input buffer` is to memorise and store data of adequate length for the purpose of network processing algorithm. Network processing algorithm realises tasks specific for a certain group of devices, e.g. for switches it provides a function of switching packets, in terms of router it is responsible for adequate tracking of packets, for firewall devices it filtrates packets. Then there is an element collecting data processed by the algorithm output buffer. Other elements of the model represent action of preparing data for sending, so the process of encapsulation of OSI model layers adequate for a certain type of a device and then the process of modulation and transmission with a use of adequate media access algorithm, analogically to IDC. Taking the model from Fig. 5 into consideration, delay introduced by network devices can be represented by the following dependency:

$$\tau_{ND} = \tau_{dem} + \tau_{dec} + \tau_{bi} + \tau_{npa} + \tau_{bo} + \tau_{enc} + \tau_{naa} + \tau_{dem} + \tau_{mod} \tag{6}$$

If in the model we determine all elements generating delays in the telecommunication system we can determine a complete delay. The complete delay connected with data transmission between IDC and ODC consistently with the proposed model is the total of delays caused by all elements included in the model from Fig. 4 and can be expressed by the below dependency:

$$\tau_{COM} = \tau_{IDC} + \tau_{ODC} + \tau_{ND} + \tau_{tm} \tag{7}$$

The proposed model of data transmission delays system allows for the analysis of occurrences connected with individual, single sources of delays in the transmission system. Such analysis facilitates the construction of devices used for measurement of data transmission delays in computer networks [7,9,11].

3 Probabilistic Modelling of Communication Delays

The probabilistic analysis of delays for measurement and control data transmission networks enables to determine the distribution of delays occurring due to factors described randomly [6,10]. They have a direct influence on the performance of the whole automation system which uses data transmission network for controlling individual elements.

Such description is the natural consequence of the fact that decisions about establishing communication in the network are made accidentally. Moreover, the probabilistic description of delays provides full information on their characteristics, therefore its usage for the purpose of network characteristics analysis can be significantly broader than in the case of deterministic description [10].

For probabilistic description of the components of the expression (8) it is best to use probability density function. Bearing in mind that delay cannot represent negative values, the probability density function of delay of transmitter access to the transmission media can be represented by the dependency below:

$$g_{\tau_{IDC}}(\tau_{IDC}) = \begin{cases} g_{IDC}(\tau_{IDC}) & for \quad \tau_{A1} \leq \tau_A \leq \tau_{A2}, where \quad \tau_A \geq 0 \\ 0 & for \quad other \end{cases} \tag{8}$$

but $g_{IDC}(\tau_{IDC})$ is a continuous and bounded function, which fulfils the following normalization condition:

$$\int_{-\infty}^{\infty} g_{IDC}(\tau_{IDC})\,d\tau_{IDC} = \int_{\tau_{A1}}^{\tau_{A2}} g_{IDC}(\tau_{IDC})\,d\tau_{IDC} \tag{9}$$

Assuming, that probability density function $g_t(\tau t)$, describing distribution of delays caused by transmission through media, is expressed by the following series:

$$g_t(\tau_t) = a_0\delta(\tau_t + \tau_{IDC} - \tau_0) + a_1\delta(\tau_t + \tau_{IDC} - \tau_1) + \ldots + a_k\delta(\tau_t + \tau_{IDC} - \tau_k) \quad (10)$$

where δ is the symbol of Dirac delta function [11] but, if we include the properties of delta function, there occurs:

$$\int_{-\infty}^{\infty} \delta(\tau_t - \tau_i)\, d\tau_t = 1 \quad (11)$$

where i is the moment number, and the number of terms of a series (10) is limited and equals $k+1$. The coefficients a_0, a_1, ..., a_k of sequence (10) have constant and non-negative values.

The dependency (10) is the effect of the analysis of occurrences happening during the transmission of information through transmission channel, which is realized with the assumption that in the case of improper transmission it is repeated so the retransmission of data occurs. The maximum number of repetitions is k. The individual terms of series (10) describe repetition of transmission, but time τ_i indicates a moment of i-number transmission repetition, $i = 1, \ldots, k$, and coefficient a_i determines probability of repetition of transmission no i. When partial delays are independent, the complete delay distribution, consistently with the formula (2), can be determined with a use of convolution of probability density function of those delays.

In that case the probability density function of complete delay is expressed by the dependency below:

$$g_{ODC}(\tau_{ODC}) = g_{IDC}(\tau_{IDC}) \otimes g_t(\tau_t) \quad (12)$$

The convolution is a linear transformation [11], so after introduction of equation (10) to the expression (12) we can present them as:

$$\begin{aligned} g_{ODC}(\tau_{ODC}) = a_0 g_{IDC}(\tau_{IDC}) \otimes g_t(\tau_t + \tau_{ODC} - \tau_0) + \\ a_1 g_{IDC}(\tau_{IDC}) \otimes g_t(\tau_t + \tau_{ODC} - \tau_1) + \ldots + \\ a_k g_{IDC}(\tau_{IDC}) \otimes g_t(\tau_t + \tau_{ODC} - \tau_k) \end{aligned} \quad (13)$$

Taking the dependency (2) into consideration we can state, that expression (13) is the sum of components, from which each constitutes a product of coefficient and convolution of delta function and function of probability delays distribution introduced by a transmitter. We can also prove, that for the component i there occurs:

$$g_{ODCi}(\tau_{ODC}) = g_{IDC}(\tau_{IDC}) \otimes \delta(\tau_{ODC} - \tau_i) = g_{IDC}(\tau_{ODC} - \tau_i) \quad (14)$$

Example

Assuming, that sequence (10) was reduced to the first two terms the dependency (14) is expressed as follows:

$$g_{ODC}(\tau_{ODC}) = a_0 g_{\tau_{IDC}}(\tau_{ODC} - \tau_0) + a_1 g_{\tau_{IDC}}(\tau_{ODC} - \tau_1) \quad (15)$$

For exemplary values: $\tau_0 = 5$, $\tau_1 = 10$, and $a_0 = 0{,}7$ and $a_1 = 0.3$ we receive:

$$g_{ODC}(\tau_{ODC}) = 0.7 g_{\tau_{IDC}}(\tau_{ODC} - 5) + 0.3 g_{\tau_{IDC}}(\tau_{ODC} - 10) \quad (16)$$

Such record of complete delays distribution means that 70 percent of data reaches the receiver at the first time, and the remaining 30 percent is transmitted during the second time. The exemplary distribution described with a series (16) for a normal communication distribution having the following parameters (0,1) is presented in Fig. 6.

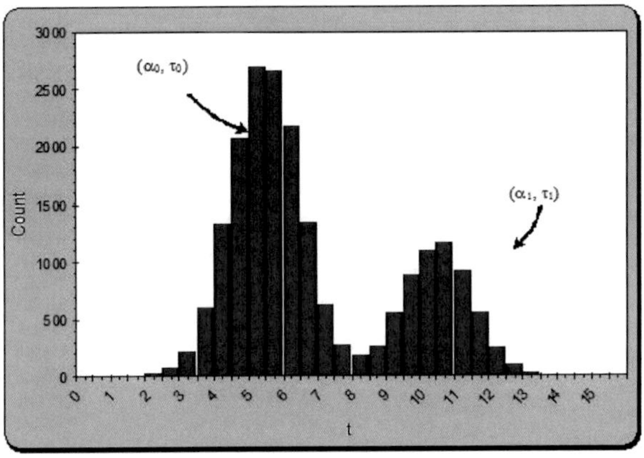

Fig. 6 Delays distribution described with a use of series (16) for example 1

The single step of the experiment involves the selection of x and y, which probability of occurrence is respectively 0.7 and 0.3. Then the number τ is sampled with a normal distribution, to which the value 5 is added, when the number x occurs or 10, when the number y occurs.

4 Experimental Verification of Delays Model in Wireless Systems

Nowadays, there are a few wireless data transmission standards available on the market. The most popular include: GSM/GPRS, Bluetooth, Wi-Fi and ZigBee. In the case of data transmission between the system of autonomous sensors dispersed in space, it seems that most often ZigBee is the most suitable protocol for the requirements of the monitoring system.

4.1 ZigBee Modules

In the wireless protocols for data transfer such as Bluetooth (IEEE 802.15.1) or Wi-Fi (IEEE 802.11), the data transmission should be ensured strictly to the chosen addressee without crosstalk and errors. In order to guarantee integrity and accuracy of the transmitted data, the protocols have to fulfil certain requirements [15].

- Media access - two nodes cannot transmit data at the same time to prevent collisions and corruption of data.
- Addressing - there is a possibility of sending message to a chosen recipient or to a few recipients without the need to broadcast it in the whole network.
- Error detection - the protocols must ensure efficient detection of incorrect transmissions.
- Confirmations and retransmissions - there is a way to inform a node transmitting data, that the data which reached the receiver are correct. In the case of failure there is a possibility of repeating the transmission for a few times.

There are many ZigBee module devices of many manufacturers with significantly different mechanical and software parameters.

4.2 XBee Modules

Among the available modules the XBee ones (Fig. 7.) have been selected, which also work on the basis of ZigBee standard. They ensure the fulfilment of all the requirements described above and construction of private wireless networks WPAN (Wireless PAN). Due to low throughput (250 kbps) they are often referred to as Low-Rate WPAN. The protocol uses the CSMA/CA mechanism to ensure transmission control. XBee modules enable construction of relatively cheap wireless networks with low power consumption. Each module ensures effective data exchange within the network with minimum power consumption, which is used only during transmission and reception of messages. XBee works with the frequency of 2.4GHz, has ports for data transmission with serial connector RS232 or RS485, analogue inputs and additional digital data inputs/outputs. Depending on the chosen software they can work as devices supporting the network layer IEEE 802.15.4, or as devices defined by the ZigBee protocol.

The XBee-PRO with increased output power of the transmitter ensuring bigger range of transmission were used in the experiment. The XBee-PRO module can work in two different modes: AT and API. The AT is the mode of see-through

Fig. 7 XBee module

transmission, in which data are only transmitted. The two modules XBee PRO practically create a wireless serial connection allowing for a simple exchange of information. In order to configure modules the AT commands are used. In the API mode, data are encapsulated into headings including, among others: destination address, source address, type of packet, control total and information on the strength of radio signal. It increases the flexibility and reliability of the system. In the case of XBee-PRO modules, elements building the network do not have to work in the same mode and this is a great advantage of those modules.

4.3 Measurement Circuit

In order to conduct verification experiments the adequate metering circuit was made. The general concept of metering circuit construction was based on the usage of two modules which send data wirelessly between each other with reference to determined time intervals. The measurement of delay was made with a use of wire coupling between the separated parts of the measurement circuit.

The task of Module 1 is to wirelessly send data supplied by serial port to Module 2, in every determined time interval. The received data are put by Module 2 at its serial output which causes signalisation of interruption in the microcontroller. Time between sending data to Module 1 and their reception from Module 2 is registered by PC.

4.4 Mesurement Circuit Consdtruted on the Basis of XBee Modules

The advantage of using XBee-PRO modules of Digi International is the possibility of direct transmission between two end devices without necessity of introducing a coordinator. It ensures simple construction of control system and software.

The aim of the measurements were to determine basic parameters of wireless transmission:

- number of properly received data bytes with reference to the total number of the sent bytes,
- communication delay understood as a time between byte sending by a microprocessor, and its reception by the second module,
- power ratio of the received radio signal,
- number of retransmitted bytes in reference to the total bytes which were sent.

Communication with module through serial connector has been broadly described in the documentation of the manufacturer [12]. Figure 8 present a connection of two modules by means of a serial interface with microcontrollers.

Each XBee-PRO module is equipped with a UART interface (Universal Asynchronous Receiver and Transmitter), and a few analog and digital inputs/outputs

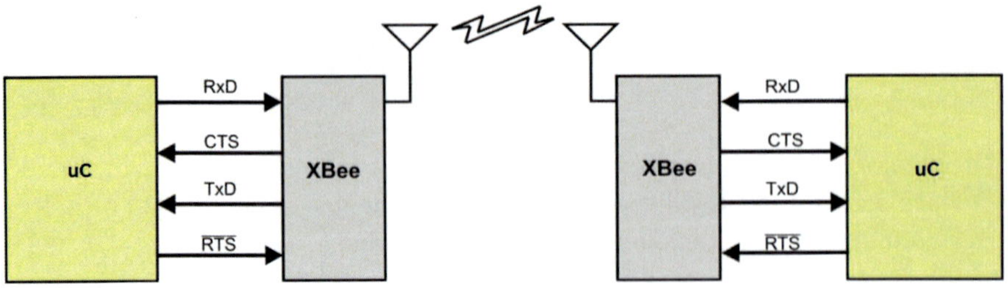

Fig. 8 Communication with XBee-PRO via UART interface

to improve its capabilities [12]. UART makes it possible to connect any microcontroller with an implemented serial bus directly to the XBee-PRO module. In order to connect the XBee-PRO module to a computer unit, the module's voltage levels have to be converted into output RS-232 voltage standards. Basic communication requires only two lines: TxD data sending and RxD data receiving. CTS and RTS confirmation signals' usage is advised while changing or updating the module's software [13]. RxD input signal should have a value of logical one while it does not transmit any data. After the start bit, eight bits of data are transmitted with the stop bit at the end. Default speed of module's transmission is 9600 bps, which means that transmission of one bit of data lasts about 1 ms.

Measurement system scheme is presented in Figure 9. Two XBee-PRO modules with their output power improved in relation to normal XBee-PRO modules are objects of the study. One of them is responsible for data transmission, and the second one for its receiving. A component used to control the performance of the measurement system is AVR ATTiny2313 microcontroller, sending data to serial input of the transmitting module. PRO modules have higher output power of radio signal as compared to traditional modules and were used due to their accessibility. During the survey, single byte transmission was used. After sending the start bit to the transmitting XBee-PRO, the microcontroller starts to measure the time, by using its internal clock. The second of the modules, implemented into the same network and working in the same radio channel, receives these data and transfers it to TxD serial bus output. Occurrence of the data start bit in the line, which is connected with interrupt (INT) output of the microcontroller, leads to a signalization of interruption and a stoppage of the clock. The period between data transmission and interruption is registered as wireless communication delay. Accuracy of time measurement in the implemented microcontroller is equal to 2 μs. The value of the measured time is transmitted from the microcontroller to a computer. Data transferred from the measurement system to PC is registered and saved as a text file. Another data byte is sent to the transmitting module after 100 ms from sending the first data byte. This value is constant and high enough not to interfere with the transmission of the last byte and possible repetitions (retransmissions) of data. Duration of the whole measurement cycle depends on system users and can be stopped at any time. Concerning the necessity to create a histogram, it is advised to obtain at least several thousands of measurement results.

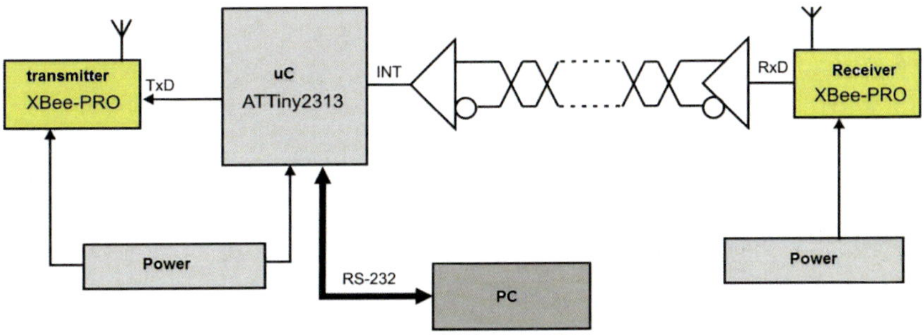

Fig. 9 A block diagram presenting the operation of the transmission delay measurement system

Symmetrical transmission was used because it is one of the most popular methods of wireless data transmission in industrial networks, and differential transmission prevents external interferences from influencing data sending on big distances. Due to technological purposes, it was necessary to use a long cable and the influence of external interferences had to be eliminated, because they could induce transmission disturbances and generate invalid data. A twisted-pair cable was used to transmit the data, with each twisted-pair additionally foiled in the screen made of foil and SF/FTP netting. The screen of the cable was connected with numerous systems, and in order to mute the interferences, ferrites were used on the cable. Screening of the measurement system module occurred to be unnecessary. Concerning big distances between the modules, they were powered separately.

A view of a prototype measurement system is presented in Figure 10.

Fig. 10 Experimental measurement system

In order to configure wireless transmission parameters (transmission channel, network addresses), a unit converting signals was used converting RS-232 standard signals into XBee-Pro acceptable logical levels, as it is presented in Figure 11.

XBee-PRO modules have an implemented data retransmission algorithm, helpful when the transmitted data did not reach the receiver [12]. After a successful transmission, the receiver sends a confirmation of ACK data reception to the transmitter.

Fig. 11 RS-232 to TTL level converter

No confirmation makes the module resend the previous data automatically. XBee-PRO modules retransmit corrupted or lost data up to 3 times. If the confirmation is not present 4 times, the data is recognized as lost. Another byte of data from the processor is transmitted to the sending XBee-PRO after such a period, to make it possible to register all possible signal retransmissions.

4.5 *Passive Interferences*

Wireless data transmission can be interfered by numerous obstacles. Because of the importance of the radio-transferred data, the number of correctly received packets by the receiver, in relation to the total number of packets sent by the transmitter, becomes an important parameter of the network. This parameter makes it possible to specify the participation of lost packets as one of possible indicators of the performance of the surveyed wireless network. The system designed and described in this publication, apart from indicating the participation of the lost packets, facilitates measurement of effective delay time of transferred packets including time used to retransmit the lost packets.

Wireless transmission interferences can be both passive and active. Passive interferences are those suppressing radio signals and not having a possibility to generate or induce electrical signals. They are, among others, all architectonical components of buildings, walls, plates, fencings made from various materials and, for example heavy rains or bushy plants. Active interferences are those generating signals, which may lead to overlapping of radio waves of a measurement system and waves generated by interfering units. As a result, distorted data (or data coming from other sources) reach the receiver.

4.6 Open Space Transmission

Assuming that the aim of the measurements is to determine complete delay distribution, when no other factors influence the transmission, function series is reduced to one formula [10]:

$$g_{ODC}(\tau_{ODC}) = a_0 g_{IDC}(\tau_{ODC} - \tau_0) \tag{17}$$

Measurements were made for a communication channel including 2 XBee modules placed in direct visibility, transferring a message on fixed data length between them and placed at a distance of 1 m from each other, as it is presented in Figure 12. No obstacles were used between the transmitter and the receiver.

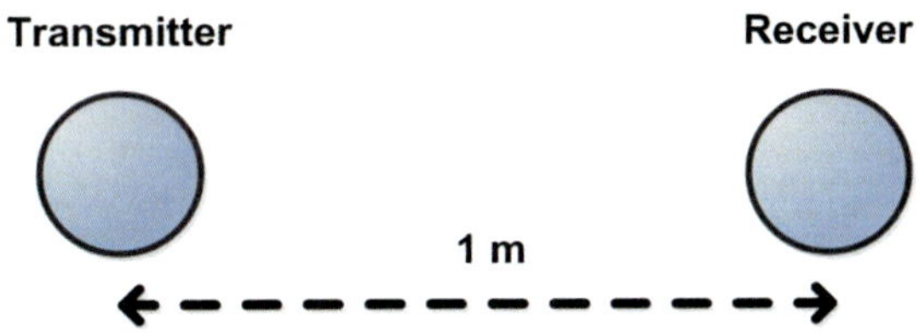

Fig. 12 Placement of modules in measurement conditions

Results were presented in the histogram (Figure 13), which presents transmission time of individual packets (number of delays - effective times of transmission in a set period). 20.000 attempts were made, and each of them was a measurement of a single transmission between two end devices, working in a point-point configuration. Each of the transmission attempts was accomplished successfully. Average propagation time is equal to 8.10 ms. There is a range of measurement results amounting to ±0.06 ms and it is a result of the properties of a wireless module

Table 1 presents a percentage value of packets received by the receiver and an average delay value (effective transmission time).

Table 1 Delay measurement results

	Number of packets received, %	Delay, ms
Direct transmission	100	8.10

On this basis, it can be assumed that the total delay distribution can be expressed according to the following formula:

$$g_{ODC}(\tau_{ODC}) = 1 g_{IDC}(\tau_{IDC} - 8.10) \tag{18}$$

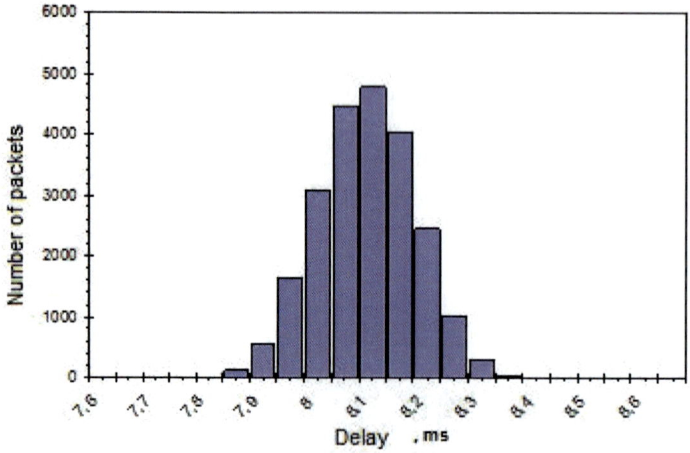

Fig. 13 The histogram showing delay results for open space transmission for XBee-PRO modules

Due to specifying the average propagation time and obtaining the histogram for work of devices in interference-free conditions, it will be possible to indicate the influence of measurement conditions on results. In order to receive more clarity of measurement results, subsequent attempts were limited only to one digit after comma while recording the delay value.

The obtained delay histogram (Figure 13) makes it possible to assume that the distribution of delays is more or less normal. The shape of the distribution is influenced only by the activity of the transmitter which precedes gaining access to the data carrier, because the time of sending each packet is the same. It means, that the distribution presented in Figure 11 (obviously, shifted by τ_0) can act as a so-called function series scheme. Therefore, measurements of this scheme should be made for the communication channel, if there are no factors influencing the transmission.

The delay histogram obtained using a simulation technique includes one part being a copy of the scheme, that is normal distribution defined by parameters (0, 0.5), and shifted by $\tau_0 = 8.10$ as it is presented in Figure 14. According to the formula (2), it is a sum of average access time to medium τ_{IDC} and transmission time τ_t.

4.7 *Transmission through a Single Wall*

An obstacle in a form of a single brick wall of a thickness of 25 cm was introduced into the measurement system. The transmitter and the receiver were located on opposite sides of the wall, with a space of 1 m between them and the wall. The

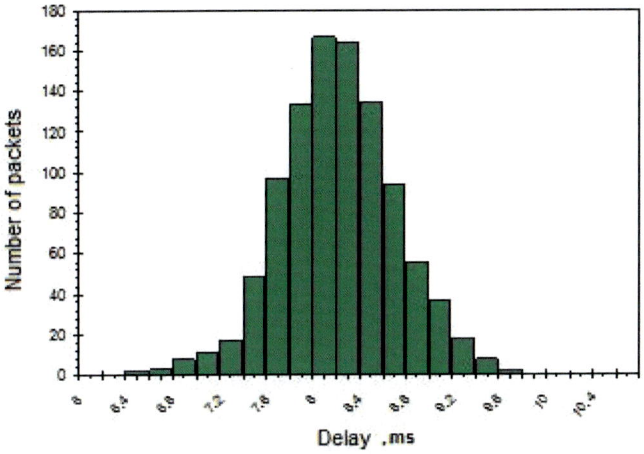

Fig. 14 The delay time histogram obtained using a simulation technique

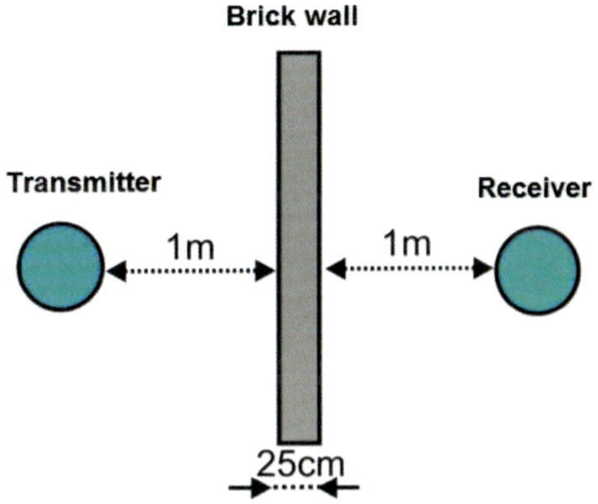

Fig. 15 Location of modules in measurement conditions

transmitter and the receiver were located in the same line perpendicular to the wall's surface (Figure 15).

The measurement consisted in transmitting a single byte of data between the modules. The number of repetitions was equal to 25.000. The obtained results were presented in a histogram (Fig.16), which presents transmission time for individual packets (the number of delays effective transmission times in a set period). Not all attempts were successfully accomplished. A small percentage of all packets sent by the transmitter was delivered to the receiver during direct transmission time. The transmitter sent several dozens of packets one more time. This repeated transmission is called the first data retransmission and is shifted in time by 10.2 ms (not visible in Figure 16 due to its legibility).

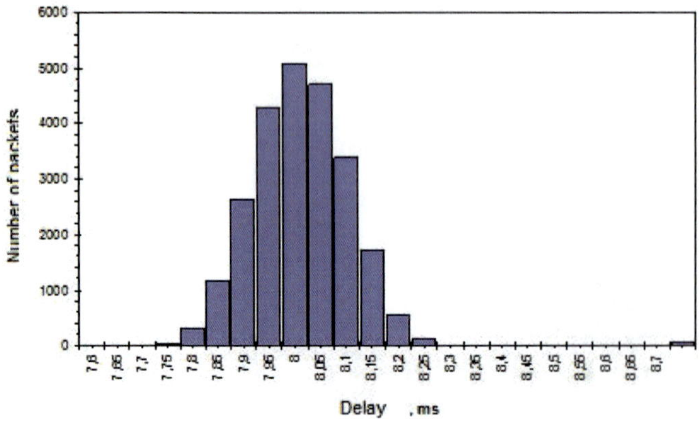

Fig. 16 The delay results histogram for XBee-PRO modules

Table 2 presents a percentage number of received packets and average delay times (effective transmission times). Despite the necessity to send some packets one more time, all packets sent by the transmitter reached the receiver at the end.

Table 2 Delay measurement results

	Number of packets received, %	Delay, ms
Direct transmission	99.8	8.10
First retransmission	0.2	10.20

The results of simulation surveys in the discussed case are presented in Figure 17. The histogram includes two parts. The first part has a scale factor of $a_0 = 0.998$, and the second one, shifted by 10.20 ms has a scale factor equal to $a_1 = 0.002$. The second copy is an equivalent of the first data retransmission of bytes, which were lost during the direct transmission.

Based on the obtained measurement and simulation results, it can be recorded that the total delay distribution can be expressed as a formula consisting of two parts:

$$g_{ODC}(\tau_{ODC}) = 0.998 g_{IDC}(\tau_{ODC} - 8.10) + 0.002 g_{IDC}(\tau_{ODC} - 10.20) \quad (19)$$

The first part concerns direct transmission and the second one - the first data retransmission.

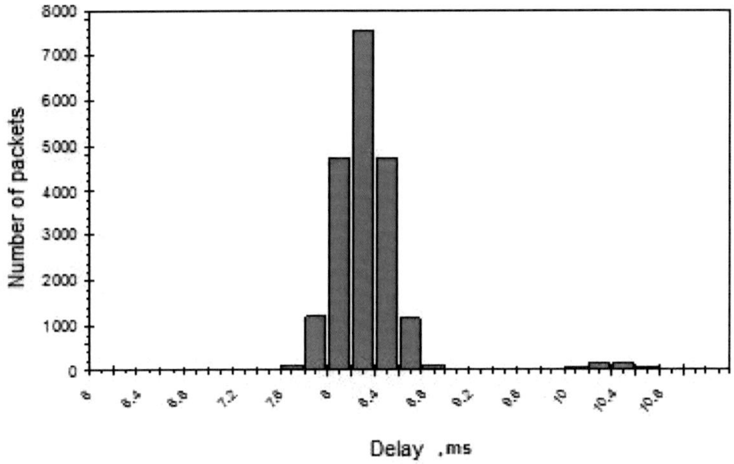

Fig. 17 The histogram obtained during a simulation of transmission for XBee-PRO modules

4.8 *Transmission through a Reinforced Brick Wall and Two Brick Walls*

Another measurement was made separating the transmitter and the receiver by three walls. One of them was a reinforced brick wall. The scheme of the measurement system is presented in Figure 18. The middle wall is the reinforced wall including metal elements. The receiver and the transmitter were located 1 m from the external walls on the same axis.

20.000 attempts of a single data byte transmission were made. The obtained histogram is presented in Figure 19. The histogram has four parts and each one is a copy of the scheme from Figure 13. However, the first copy has a scale factor equal to $a_0 = 0.82$, the second one has $a_1 = 0.16$, the third $a_2 = 0.03$ and the fourth $a_3 = 0.01$. Consecutive copies result from other data retransmissions from the transmitter to the receiver. Time shifts of all copies were gathered and presented in Table 3. Based on the gathered data it may be assumed that time shift between the copies is about 2 ms.

Figure 19 shows that in some environmental conditions, transmission of a single byte can be repeated up to three times. Some data that did not reach the destination after the first retransmission were sent again after about 2 ms. The manufacturer of XBee-PRO modules predicted a maximum of 3 retransmissions. That is why, up to four copies of a basic scheme can occur. Packets, which were not received after three retransmissions, are considered lost and are not transmitted again. Such packets are shown in Figure 17 as the last single bar, marked with inf symbol.

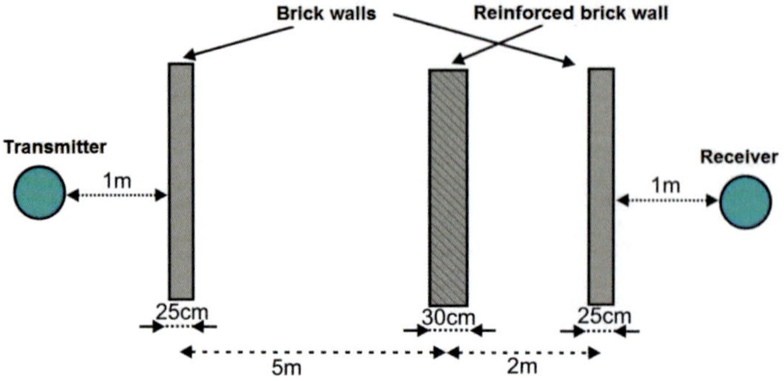

Fig. 18 Measurement system scheme

The histogram of delays obtained using a simulation technique, presented in Figure 20, was created making 10.000 attempts. It includes four parts, which are copies of the scheme, that is a normal distribution characterized by (0, 0.5) parameters and shifted by proper values of delay time.

Table 3 Delay measurement results

	Number of packets received, %	Delay, ms
Direct transmission	82	8.10
First retransmission	9	10.20
Second retransmission	3	12.40
Third retransmission	1	14.60
Lost packets	5	∞

Based on the obtained results it can be assumed that the complete delay distribution for XBee modules in a given case can be presented by the following formula:

$$\begin{aligned} g_{ODC}(\tau_{ODC}) = 0.82 g_{IDC}(\tau_{ODC} - 8.10) + 0.09 g_{IDC}(\tau_{ODC} - 10.20) + \\ 0.03 g_{IDC}(\tau_{ODC} - 12.40) + 0.01 g_{IDC}(\tau_{ODC} - 14.60) + \\ 0.05 g_{IDC}(\tau_{ODC} - \tau_1) \end{aligned} \tag{20}$$

Providing that

$$\tau_1 \to \infty \tag{21}$$

the last element of the formula (20) means that 0.05 of packets has an infinitely high delay that is they will never reach the destination.

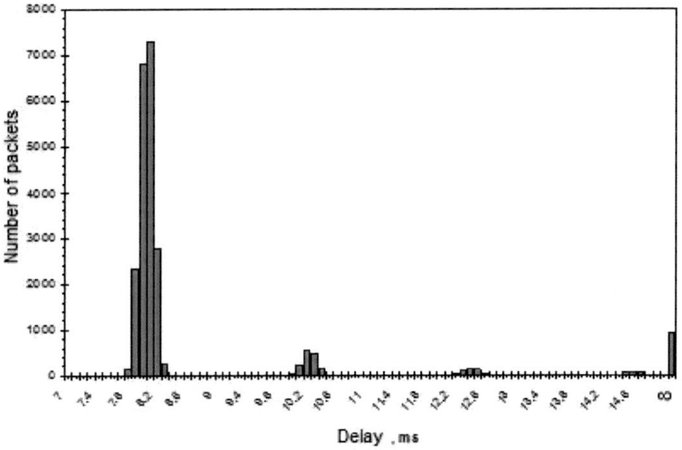

Fig. 19 The histogram of delay results for transmission through two brick walls and one reinforced brick wall for CBee modules

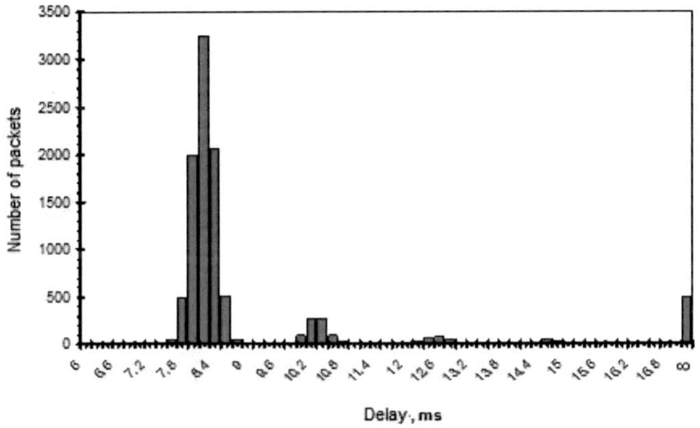

Fig. 20 The histogram created for XBee modules with transmission through two brick walls and one reinforced brick wall

5 Summary

The paper is an attempt to analyse one of data transmission parameters, namely, delays. A general delay model for the whole transmission route was proposed. Furthermore, a probabilistic analysis of this parameter was provided. Experimental verification of the hypotheses was made by means of experimental system, which uses wireless communication in the ZigBee standard. Moreover, the influence of external factors, such as architectonic elements of buildings and interior elements were presented, showing their interference with radio waves, which results in a decrease in wireless transmission quality, loss of data, as well as transmission delays.

Data taken from literature show that a number of lost packets is frequently measured, but delays in wireless and cable-based networks are rarely surveyed. There is no appropriate mathematical model to describe these delays.

This is why a new way of describing communication delays was proposed. It is based on probabilistic description of time necessary to obtain access to data source and use of a series of time-based delta functions to describe the transmission time between the transmitter and the receiver.

In order to confirm that this model is applicable for description of communication delays, a series of experiments was made, which used wireless communication modules based on IEEE 802.15.4 protocol. The experiments were concerned measuring delays between the transmitter and the receiver in different environments. The results of the surveys confirm the hypothesis that it is possible to use a characterized mathematical apparatus to describe time-related phenomena in wireless networks based on ZigBee standard.

References

1. Kielan, P.: Aspekty sterowania urządzeniami mechatronicznymi przez Internet - analiza protokołu TCP/IP. Rozprawa doktorska, Gliwice (2011)
2. Peterson, L.L., Dave, B.S.: Computer networks. A system approach. Morgan Kaufmann (2011)
3. Tanenbaum, A.S., Wetherall, D.J.: Computer networks. Pearson (2011)
4. Hassan, M., Jain, R.: High Performance TCP/IP Networking: Concepts, Issues, and Solutions. Prentice-Hall (2003)
5. Michta, E.: Modele komunikacyjne sieciowego systemu pomiarowo-sterującego. Wydawnictwo Politechniki Zielonogórskiej, Zielona Góra (2000)
6. Jakubiec, J.: Błędy i niepewności danych w systemie pomiarowo-sterującym. Wydawnictwo Politechniki Śląskiej, Gliwice (2010)
7. Topór-Kamiński, T., Żurkowski, R., Grygiel, M.: Selected methods of measuring the delay in data transmission systems with wireless network interfaces. Acta Physica Polonica A 120, 748–754 (2011)
8. Topór-Kamiński, T., Żurkowski, R.: Pomiar opóźnień transmisji danych wnoszonych przez urządzenia klasy UTM. In: Materiały VIII Konferencji Naukowej nt. Systemy Pomiarowe w Badaniach Naukowych i w Przemyśle, Łagów, pp. 171–174. Oficyna Wydawnicza Uniwersytetu Zielonogórskiego (2010)
9. Grygiel M., Żurkowski, R.: System eksperymentalny do pomiaru opóźnień transmisji w standardzie ZigBee 802.15.4, Metrologia dziś i jutro. Wydział Elektrotechniki i Automatyki. Politechnika Gdańska (2009)
10. Jakubiec, J., Krupanek, B., Grygiel, M.: Probabilistyczne modelowanie opóźnień komunikacyjnych w sieciach komputerowych. Zeszyty Naukowe Politechniki Śląskiej, seria Elektryka, z.216, Gliwice (2012)
11. Papoulis, A., Unnikrishna Pillai, S.: Probability, random variables and stochastic processes. McGrawHill (2002)
12. `http://www.digi.com/`

13. Tokarski, J., Bogacz, R., Krupanek, B.: Mikroprocesorowy system do pomiaru parametrów transmisji sieci bezprzewodowych zbudowanych w oparciu o moduły Xbee. Metrologia dziś i jutro, 321–336 (2011)
14. 802.15.4 IEEE Standard for Information technology - Telecommunications and information Exchange between systems - Local and metropolitan area networks - Specific requirements. Part 15.4: Wireless Medium Access Control (MAC) and Physical Layer (PHY) Specifications for Low-Rate Wireless Personal Area Networks (LR-WPAN), New York (2003)
15. ZigBee Specification v1.0. ZigBee Document 053473r00, ZigBee Alliance, San Ramon (2005)
16. Guide to the Expression of Uncertainty in Measurement, ISO/IEC/OIML/BIPM (1992, 1995)

Author Index